PENGUIN BOOKS

HIGH PRICE

'A hard-hitting attack on current drug policy by a neuroscientist who grew up on the streets of one of Miami's toughest neighbourhoods. An eye-opening, absorbing, complex story of scientific achievement in the face of overwhelming odds'
Kirkus Reviews

'Carl Hart is a peerless contributor to America's discourse on drug policy. He combines the rigour of a scientist with the street smarts of a former hustler. The result is *High Price*, a fiercely uncompromising look at public health, and an inspiring personal journey'
Eugene Jarecki, award-winning author and documentary filmmaker

'Life stories are powerful tools in clarifying paths to success and failure and motivating people to envision and pursue their dreams. Dr. Hart shares his story of growing up in Miami and the circuitous route he took to becoming a professor at an Ivy League university. In a unique and innovative approach, he recounts his early adulthood as a black man growing up in America and his career as a substance-abuse research scientist, in conjunction with telling the story of drug use and policy in America . . . *High Price* is highly edifying'
Lula Beatty, former director of the National Institute on Drug Abuse, and senior director at the American Psychological Association

D1043863

ABOUT THE AUTHOR

Carl Hart is an associate professor in the departments of Psychology and Psychiatry at Columbia University. He is also a research scientist in the Division of Substance Abuse at the New York State Psychiatric Institute. Dr. Hart is a member of the National Advisory Council on Drug Abuse and on the board of directors of the College on Problems of Drug Dependence and the Drug Policy Alliance. A native of Miami, Florida, Dr. Hart received his BS in psychology at the University of Maryland and his MS and PhD in experimental psychology and neuroscience at the University of Wyoming. He lives in New York City.

www.highpricethebook.com

HIGH PRICE

**Drugs, Neuroscience and
Discovering Myself**

Dr. Carl Hart

PENGUIN BOOKS

PENGUIN BOOKS

Published by the Penguin Group
Penguin Books Ltd, 80 Strand, London WC2R 0RL, England
Penguin Group (USA) Inc., 375 Hudson Street, New York, New York 10014, USA
Penguin Group (Canada), 90 Eglinton Avenue East, Suite 700, Toronto, Ontario, Canada M4P 2Y3
(a division of Pearson Penguin Canada Inc.)
Penguin Ireland, 25 St Stephen's Green, Dublin 2, Ireland (a division of Penguin Books Ltd)
Penguin Group (Australia), 707 Collins Street, Melbourne, Victoria 3008, Australia
(a division of Pearson Australia Group Pty Ltd)
Penguin Books India Pvt Ltd, 11 Community Centre, Panchsheel Park, New Delhi – 110 017, India
Penguin Group (NZ), 67 Apollo Drive, Rosedale, Auckland 0632, New Zealand
(a division of Pearson New Zealand Ltd)
Penguin Books (South Africa) (Pty) Ltd, Block D, Rosebank Office Park,
181 Jan Smuts Avenue, Parktown North, Gauteng 2193, South Africa

Penguin Books Ltd, Registered Offices: 80 Strand, London WC2R 0RL, England

www.penguin.com

First published in the United States of America by HarperCollins 2013
First published in Great Britain in Penguin Books 2013
001

Grateful acknowledgement is made for permission to reproduce from the following:
Excerpt from 'This Be the Verse' from *The Complete Poems of Philip Larkin* by Philip Larkin, edited
by Archie Burnett. Copyright © 2012 by the Estate of Philip Larkin. Reprinted by permission
of Farrar, Straus and Giroux, LLC.
'The Revolution Will Not Be Televised,' written by Gil Scott-Heron. Used by permission of
Bienstock Publishing Company.

All photographs are courtesy of the author.
Figure 1 on page 160 and figure 2 on page 290 are courtesy of the author.

Printed in Great Britain by Clays Ltd, St Ives plc

ISBN: 978-0-670-91974-1

www.greenpenguin.co.uk

ALWAYS LEARNING   **PEARSON**

FOR DAMON AND MALAKAI

Intellectuals . . . who have had the courage to voice their opposition have often paid a very high price.

—TAHAR BEN JELLOUN

The thing that makes you exceptional, if you are at all, is inevitably that which must also make you lonely.

—LORRAINE HANSBERRY

CONTENTS

AUTHOR'S NOTE

'm frequently asked why I wrote *this* book, a book that reveals many deeply personal details about my life. After all, I'm an academic neuropsychopharmacologist who is trained to conduct research and teach a select group of students about drugs, behavior, and the brain. And there are few things that I value as much as my privacy. So, I certainly didn't write the book because I thought people should know more about my private life—the vast amount of personal information revealed within these pages causes me a great deal of anxiety; nor did I write it to advocate illegal drug use—that would be a colossal waste of my training, skills, and talents. Currently, there are more than 20 million Americans who use illegal drugs regularly. I think it's clear that I'm not needed as an advocate.

The primary reason that I wrote this book was to show the public how the emotional hysteria that stems from misinformation related to illegal drugs obfuscates the real problems faced by marginalized people. This also contributes to gross misuses of limited public resources. To shed light on the relevant issues—

including maladaptive human behaviors and misguided public policies—I use real-life examples, mainly from my own life. I hope this will help the reader to learn by example and then generalize more broadly. But I also recognize that inaccurate conclusions can be easily made if personal anecdotes alone are used. Thus, in addition to real-life examples, throughout the book I used scientific knowledge of the human mind, brain, and behavior in an effort to decrease the likelihood of the reader drawing inappropriate conclusions.

In an attempt to be as accurate as possible, I visited relatives and friends and recorded what they had to say. Some of these individuals' names have been changed in an effort to protect their privacy. After absorbing the information I'd learned from meeting with them, I'd meet with the writer Maia Szalavitz, who helped me to put together a narrative that I thought would be interesting and digestible for a general audience. I gratefully acknowledge her assistance in explaining complex scientific findings and principles to a general readership, but I take full responsibility for any inaccuracy that may have resulted from oversimplification of complicated material.

It is my hope that after reading this book, you will be less likely to think about drugs in magical or evil terms that have no foundation in real evidence. As you will see in these pages, this has led to a situation where there is an unreasonable goal of eliminating illegal drug use at any cost to marginalized groups. Instead, I hope you, the reader, will come away with the ability to think more objectively and critically about the multitude of issues that come along with illegal drug use, and will understand that by applying what we have learned about human behavior, we can change it.

PROLOGUE

*The paradox of education is precisely this—that as one begins to be-
come conscious, one begins to examine the society in which he is being
educated.*

—JAMES BALDWIN

The straight glass pipe filled with ethereal white smoke. It was thick enough to see that it could be a good hit, but it still had the wispy quality that distinguishes crack cocaine smoke from cigarette or marijuana smoke. The smoker was thirty-nine, a black man who worked as a street bookseller. He closed his eyes and lay back in the battered leather office chair, holding his breath to keep the drug in his lungs as long as possible. Eventually, he exhaled, a serene smile on his face, his eyes closed to savor the bliss.

About fifteen minutes later, the computer signaled that another hit was available.

"No, thanks, doc," he said, raising his left hand slightly. He hit the space bar on the Mac in the way that he'd been trained to press to signal his choice.

Although I couldn't know for sure whether he was getting cocaine or placebo, I knew the experiment was going well. Here was a middle-aged brother, someone most people would label a "crackhead," a guy who smoked rock at least four to five times a week, just saying no to a legal hit of what had a good chance of being 100 percent pure pharmaceutical-grade cocaine. In the

movie version, he would have been demanding more within seconds of his first hit, bug-eyed and threatening—or pleading and
desperate.

Nonetheless, he'd just calmly turned it down because he preferred to receive five dollars in cash instead. He'd sampled the
dose of cocaine earlier in the session: he knew what he would get
for his money. At five dollars for what I later learned was a low
dose of real crack cocaine, he preferred the cash.

Meanwhile, there I was, another black man, raised in one of
the roughest neighborhoods of Miami, who might just as easily
have wound up selling cocaine on the street. Instead, I was wearing a white lab coat and being funded by grants from the federal government to provide cocaine as part of my research into
understanding the real effects of drugs on behavior and physiology. The year was 1999.

In this particular experiment, I was trying to understand
how crack cocaine users would respond when presented with
a choice between the drug and an "alternative reinforcer"—or
another type of reward, in this case, cash money. Would anything else seem valuable to them? In a calm laboratory setting,
where the participants lived in a locked ward and had a chance to
earn more than they usually could on the street, would they take
every dose of crack, even small ones, or would they be selective
about getting high? Would merchandise vouchers be as effective as cash in altering their behavior? What would affect their
choices?

Before I'd become a researcher, these weren't even questions
that I would think to ask. These were drug addicts, I would have
said. No matter what, they'd do anything to get to take as much
drugs as often as possible. I thought of them in the disparaging
ways I'd seen them depicted in films like *New Jack City* and
Jungle Fever and in songs like Public Enemy's "Night of the Liv-

ing Baseheads." I'd seen some of my cousins become shells of their former selves and had blamed crack cocaine. Back then I believed that drug users could never make rational choices, especially about their drug use, because their brains had been altered or damaged by drugs.

And the research participants I studied should have been especially driven to use drugs. They were experienced and committed crack cocaine users, who typically spent between $100 and $500 a week on it. We deliberately recruited individuals who were not seeking treatment, because we felt that it would be unethical to give cocaine to someone who had expressed an interest in quitting.

The bookseller was seated in a small, bare chamber at Columbia-Presbyterian Hospital (now New York–Presbyterian) in upper Manhattan; his cocaine pipe had been lit by a nurse at his side with a lighter, who also helped monitor his vital signs during the research. I was watching him and several others in similar rooms through a one-way mirror; they knew we were observing them. And over and over, these drug users continued to defy conventional expectations.

Not one of them crawled on the floor, picking up random white particles and trying to smoke them. Not one was ranting or raving. No one was begging for more, either—and absolutely none of the cocaine users I studied ever became violent. I was getting similar results with methamphetamine users. They, too, defied stereotypes. The staff on the ward where my drug study participants lived for several weeks of tests couldn't even distinguish them from others who were there for studies on far less stigmatized conditions like heart disease and diabetes.

To me, by that point in my career, their myth-busting behavior was no longer a surprise—no matter how odd and unlikely it may seem to many Americans raised on Drug Abuse Resistance

Education (DARE) antidrug programs and "This is your brain on drugs" TV commercials. My participants' responses—and those in the dozens of other studies we'd already run, as well as studies by other researchers around the country—had begun to expose important truths. Not just about crack cocaine and about addiction, but about the way the brain works and the way that pleasure affects human behavior. Not just about drugs, but about the way science works and about what we can learn when we apply rigorous scientific methods. This research was beginning to reveal what lies behind choice and decision-making in general and how, even when affected by drugs, it is influenced powerfully by other factors as well.

These experiments were potentially controversial, of course: the tabloids could have described me as a "taxpayer-funded pusher, giving 'crackheads' and 'meth-monsters' what they want."

Nevertheless, I tried to keep the sensational stuff hidden in the mantle and cold language of science in my scholarly publications. I'd published dozens of papers in important journals, had been awarded prestigious fellowships and competitive grants to conduct research, and had been invited to join influential scientific committees. I cowrote a respected textbook that became the number-one text used to teach college students about drugs; I won awards for my teaching at Columbia University. But throughout my career I mainly tried to avoid controversy, fearing it might derail me from conducting the work I so loved.

Eventually, I realized that I could no longer stay silent. Much of what we are doing in terms of drug education, treatment, and public policy is inconsistent with scientific data. In order to come to terms with what I have seen in the lab and read in the scientific literature, there is nothing else to do but speak out. Using empirical data, not just personal anecdotes or speculation, I have to discuss the implications of my work outside the insulated and cautious scientific journals, which were my normal métier. Because basically, most of what we think we know about drugs, addiction, and choice is wrong. And my work—and my life—shows why.

As I monitored the people I was studying, I began to think about what had brought each of us to such different places. Why was I the one in the white coat—and not the crack cocaine smoker in the cubicle? What made us different? How did I escape the distressed neighborhoods I grew up in—and the adult lives marked by drugs, prison, violent death, and chaos that so many of my family and childhood friends have had? Why did I instead become a psychology professor at Columbia, specializing in neuropsychopharmacology? What allowed me to make such different choices?

These questions weighed on me even more heavily later in the year as I continued to conduct these experiments. Sometimes, while I watched the drug users contemplate whether to take another dose, I couldn't help thinking about some of the choices I'd made during my youth. Marvin Gaye's lyric from "Trouble Man" would run through my head, especially the lines about growing up under difficult circumstances, but eventually turning the tables to succeed. Usually, I tried to keep my past far behind me. But that part of my life had been called to my attention in an unavoidable and shocking way that spring.

Early one morning in March 2000, I was awakened by a loud banging on the door of my Bronx apartment. It was about 6 a.m.; I was in bed with my wife. We had a young son, Damon, who was about to turn five. Several months earlier, I had been promoted to assistant professor at Columbia. Life was good. As we say back home, I was feeling myself. But I also knew that word of my success had hit the streets of South Florida. Indeed, I'd recently received what I thought was an absurd letter from a

Florida court claiming that I was the father of a sixteen-year-old boy. The pounding became more insistent.

When I opened the door, I was met by a thick-necked white guy wearing an undersized suit and displaying a badge. He handed me some official paperwork and instructed me to appear before a judge. As it turned out, the boy's mother had actually gone ahead and filed a paternity suit. I'm embarrassed to say that I didn't even know her last name. But, in the fall of 1982, when I was fifteen and she was sixteen, we'd had a one-night stand. It started to come to me as I thought back; soon I had a vague memory of her signaling me to sneak in through her window to avoid alerting her mother that she had a visitor.

As the DNA test ultimately confirmed, I'd gotten her pregnant that night. For the next two years, prior to joining the U.S. Air Force, I'd lived in and around the Carol City neighborhood of Miami (known to hip-hop fans as the gun- and drug-filled home of rapper Rick Ross and his Carol City Cartel), but she had never even mentioned the possibility to me that I was the father of her baby boy. And I never even thought to ask, because I had engaged in this type of behavior in the past without notice-able consequences.

But that's the abrupt way I discovered that I had a son I didn't know—one who was being raised in the place I'd tried so hard to escape; yet another fatherless black child of a teenage mother. At first, I was enraged, horrified, and embarrassed. I thought I had at least avoided making that mistake. Here I was doing the best I could to raise the child I knew I had in a middle-class, two-parent family. I couldn't believe it. I didn't know what to do. Once I got over my initial shock, I was appalled to think about what it must have been like for my son to grow up without ever knowing his father. It really got me thinking about how I'd managed to thrive despite lacking those advantages.

I'd wanted to teach my children everything I hadn't known as I grew up with a struggling single mother, surrounded by people whose lives were limited by their own lack of knowledge. I wanted them to go to good schools, to know how to negotiate the potential pitfalls of being black in the United States, to not have to live and die by whether they were considered "man" enough on the street. I also wanted to illustrate by my own example that bad experiences like those I had as a child aren't the defining factor in being authentically black.

Now I had learned that one of my own children—a boy, whose name I learned was Tobias—had grown up for sixteen years in the same way I had, but without any of the hard-earned knowledge I could now offer.

Later, I'd discover as well that he'd taken the very path I feared most. He had dropped out of high school and fathered several children with different women. He had sold drugs and allegedly shot someone. What could I tell my sons about how I'd escaped from the streets? Could my experience and knowledge help change Tobias's downward trajectory? How did I really manage to go from being one of the black kids in the auxiliary trailer for those with "learning difficulties" in elementary school to being an Ivy League professor?

Though I now regret much of this behavior, like my new-found son I'd sold drugs, I'd carried guns. I'd had my share of fun with the ladies. I'd deejayed in the skating rinks and gyms of Miami performing with rappers like Run-DMC and Luther Campbell in their early gigs, ducking when people started shooting. I'd seen the aftermath of what the police call a "drug-related" homicide up close for the first time when I was just twelve years old; I lost my first friend to gun violence as part of the same chain of events. Indeed, my cousins Michael and Anthony had stolen from their own mother, and I had attrib-

uted this abhorrent behavior to their "crack cocaine addictions."
I saw what happened as crack first took hold in Miami's poorest
black communities. Falling for media interpretations and street
myths about all of these experiences had originally misled and
misdirected me. Some of that, as we shall see, may ironically
have helped me at certain times. But more often, it was a distraction, one that prevented me and so many others in my community from learning how to think critically.

So how could I now in good conscience study this scourge
of a drug, even offer it to my own people in the laboratory? In
the grand scheme of things, what was really so different between
what I was doing in my research and what was likely to get
Tobias arrested on the street?

The answers lie in my story and the science, which reveal the
untold truth about the real effects of drugs and the choices we
make about them as a society. By exploring how these myths
and social forces shaped my childhood and career, we can strip
away the misinformation that actually drives so-called drug epi-
demics and leads us to take actions that harm the people and
communities we presumably intend to help.

CHAPTER 1

Where I Come From

This nation has always struggled with how it was going to deal with poor people and people of color. . . . We've had the war on poverty that never really got into waging a real war on poverty.

—MAXINE WATERS

The sounds were what got to me: my father shouting, "I'll kill yo ass"; my mother shrieking; the sickening thump of flesh hitting flesh, hard. I had been playing board games—probably Operation or something like that—with three of my sisters in the bedroom I shared with my youngest brother, Ray. He was three, too young to play, but I was watching him, keeping him out of trouble. The fierce Miami sun was setting and we could tell the fighting was getting worse because my parents had moved from their bedroom, where they tried to keep things private, into the living room, where anything went.

It was a Friday or Saturday night and I was six years old.

Soon we could hear large objects being thrown against the walls, glass shattering, long, piercing screams. I had known it was going to be a bad night when my oldest sister, Jackie, left to go home. Then thirteen, Jackie was the child of my mother's previous partner, born when my mother was eighteen, before my parents had met and gotten married. She lived with Grand-mama, as we called our maternal grandmother, but during her frequent visits with us, she was sometimes able to prevent my parents from attacking each other.

Not this time. Maybe she had sensed what was coming. It was worse than ever—even worse than some of the other times when the neighbors had called the police. In 1972—long before Farrah Fawcett's *The Burning Bed* and O.J. and Nicole—the courts were reluctant to prosecute domestic violence cases, in part because they didn't want to incarcerate the family's primary wage earner, which might have left the wife and children desti-tute. As a result, domestic violence was a tolerated behavior and was not limited to black families. The police would eventually come and they would talk to my father. Sometimes they would tell him to go away for a bit to cool off, but they never arrested him. They saw it as private, something between a man and his woman. I felt relieved when they broke things up, but I didn't understand why the fights never stopped.

My sisters whispered to each other for a split second, then took the youngest ones by the hand and pulled us through the living room into the yard. Patricia, then nine years old, stayed behind. She often tried to play peacemaker like her big sister Jackie. The terrifying screams and crashes continued. Ten-year-old Beverly and seven-year-old Joyce tried to get me out as quickly as possible but I still saw my father hit my mother with a hammer. The glass coffee table that was usually in front of the couch was shattered. Shards of glass were everywhere. The ceramic lion that I once got grounded for accidentally dropping wielded its claws in empty menace by the front door.

I froze but my sisters dragged me along. The poster-sized photos of Martin Luther King and JFK on the living room wall looked dead in their frames. As we ran out, I looked back to see my mother collapse, bleeding, at the door that opened from the living room into the yard. What I remember most is horror. The memories themselves are disjointed, as if reflected in the splintered glass.

"My mama's dead!" one of the girls screamed. "My mama's dead!"

"Carl done killed my mama," another sister said. In my family, we never called our father Dad or Daddy, just used his first name, for reasons now lost to family history.

"Carl done caught her in the head with a hammer!" Beverly, my third-oldest sister, shrieked.

Someone, probably our next-door neighbor who'd made these types of calls before, dialed 911. An ambulance arrived and took my mother away to the hospital. At some point, her father, whom we called Pop, came to collect us and took us to our maternal grandmother's house. But no one told me how my mother was or anything about what was going on. And it didn't occur to me to ask: in our family, you didn't really raise those kinds of questions. I learned that she was alive only when she turned up a few days later, with blackened eyes and a bandage on one arm.

No crack cocaine was involved in my family life. That drug would not appear on the scene until the 1980s and I was born in 1966. There wasn't any powder cocaine or heroin, either. Alcohol, however, was definitely part of the chaos. My father never drank during the week. But weekends were his time to let go, to make up for the social and cultural isolation of his work as a warehouse manager. At the time, he was one of two black employees at his company and the only one in management. His whiskey with Coke chasers were his reward—and Friday nights were his time to hang out on the corner with his friends.

My brother Ray (right) and me
on Easter Sunday 1972.

All of my parents' worst fights took place on weekends. Most were either Friday or Saturday night when he was drunk, or Sunday when he was hung over. As a result, unlike typical school-age children, my siblings and I dreaded weekends. My mom, Mary, would drink when people around her drank, but drinking wasn't a specific pursuit for her the way it was for my father. She imbibed for social reasons while he drank to get intoxicated and experience the disinhibiting effects of alcohol.

But although alcohol was involved, I now know it wasn't the real root of our problems. As a scientist, I have learned to be skeptical about the causes attributed to the difficulties that my family faced, living first in a working-class and later in a poor community. Simple factors like drinking or drugs are rarely the whole story. Indeed, as we know from experience with alcohol, drinking itself isn't a problem for most people who do it. As we will see, the same is true for illegal drugs, even those we have learned to fear, like crack cocaine and heroin.

While I could tell my story without highlighting what I've learned about these issues, that would merely perpetuate the misinterpretations that misguide our current thinking. To truly understand where I came from, you have to understand where I wound up—and how mistaken ideas about drugs, addiction, and race distort the way we see lives like mine and therefore, how society addresses these questions.

First, in order to understand the nature of influences like alcohol and illegal drugs, we need to carefully define the real nature of the problems related to them. Knowing that someone uses a drug, even regularly, does not tell us that he or she is "addicted." It doesn't even mean that the person has a drug problem.

To meet the most widely accepted definition of addiction—the one in psychiatry's *Diagnostic and Statistical Manual of Mental Disorders*, or *DSM*—a person's drug use must interfere with important life functions like parenting, work, and intimate relationships. The use must continue despite ongoing negative consequences, take up a great deal of time and mental energy, and persist in the face of repeated attempts to stop or cut back. It may also include the experience of needing more of the drug to get the same effect (tolerance) and suffering withdrawal symptoms if use suddenly ceases.

But more than 75 percent of drug users—whether they use alcohol, prescription medications, or illegal drugs—do not have this problem.[1] Indeed, research shows repeatedly that such issues affect only 10–25 percent of those who try even the most stigmatized drugs, like heroin and crack. When I talk about addiction in this book, I always mean problematic use of this sort that interferes with functioning—not just ingesting a substance regularly.

So why is our image of the illegal drug user so negative? Why do we think that drug use is addiction and that degradation is the primary result of taking drugs? Why do we so readily blame illicit drugs for social problems like crime and domestic violence?

Part of what I want to do here is look critically at why we see drugs and their users the way we do, the role racial politics has played in this perception, and how that has led to drug-fighting tactics that have been especially counterproductive in poor communities. I want to examine the way we ascribe causes to people's actions and fail to acknowledge the complexity of the influences that guide us on the paths we take through life. I want to explore the research data that is often used to back the claims that people make about drugs, addiction, and racism and reveal what it can and cannot tell us about these issues. By looking at how these issues affected my own life, I hope to help you see how mistaken ideas impede attempts to improve drug education and policy.

However, before proceeding I also need to clearly define one more term: *racism*. So many people have misused and diluted the term that its perniciousness gets lost. Racism is the belief that social and cultural differences between groups are inherited and immutable, making some groups inalterably superior to others. While these ideas are bad enough when lodged in the minds of individuals, the most harm is done when they shape institutional behavior, for example, that of schools, the criminal justice system, and media. Institutionalized racism is often much more insidious and difficult to address than the racism of lone individuals, because there's no specific villain to blame and institutional leaders can easily point to token responses or delay meaningful action indefinitely. I hope to shed some light on how that works here—but I never want to give the impression that I am overemphasizing its force or exaggerating when I use that word. I mean precisely the role that the belief in innate racial inferiority plays in shaping group behavior.

By looking closely at all these factors, I hope to understand what forces held me back in my early educational experiences and what pushed me forward; what early influences were positive and which were negative; what happened by chance and what happened by choice; and what helps and harms children who face the same kind of chaos that I did. What allowed me—but not many of my family members and friends—to escape chronic unemployment and poverty, and to avoid prison? Can I give my own children the tools that worked for me? How do drugs and other sources of pleasure interact with cultural and environmental factors like institutional racism and economic deprivation?

It became clear to me quite early in life that things are often very different from the way they seem on the surface; that people present very different faces to the world at work, in church, at home, and with those they love most. That complexity is also found in some interpretations of research data. As citizens in a society where there are many people with varying agendas trying to wrap themselves in the cloak of science, it's important to know how to think critically about information that is presented as scientific, because sometimes even the most thoughtful people can be duped.

I want to explore with you what I've learned, especially the importance of empirical evidence—that is, evidence that comes directly from experiments or measurable observations—in understanding issues like drugs and addiction. Importantly, such evidence is reliable and experiments are designed to avoid the bias that can come from looking at one or two cases that may not be typical. The opposite of empirical evidence is anecdotal information, which cannot tell us whether the stories told are outliers or are ordinary cases. Many people rely on personal anecdotes about drug experiences to try to understand what drugs do or don't do, as if they are representative cases or scientific data. They are not. It is easy to get bamboozled if you do not have specific tools for critical thinking, such as understanding different types of evidence and argument. I'll share these tools throughout this book.

All that said, what I do know for sure is that in my neighborhood, long before crack cocaine was introduced, many families were already being torn apart by institutional racism, poverty, and other forces. In his classic book *World of Our Fathers*, Irving Howe reminded us that the pathology seen in neighborhoods like mine was not unique to black communities. Many early immigrant Jewish families from Eastern Europe were disrupted by hostilities faced from other groups and poverty, which required family members to work different schedules and made it impossible for them to spend time together. Some were required to conceal or abandon their religious beliefs and customs in order to obtain marginal employment. As a result, it's not surprising that many early Jewish immigrant communities were plagued by crime, men abandoning their wives, prostitution, juvenile delinquency, and so forth. When these things happened in my neighborhood in the 1980s and 1990s, crack was blamed. For example, although crack is often blamed for child abandonment and neglect and for grandmothers being forced to raise a second generation of children, all those things happened in my family well before crack hit the streets.

My own mother, who was never an alcoholic or addict of any type, left me and the rest of her children to be raised by her relatives for more than two years during my early childhood. Some of my siblings were not raised by her at all. My maternal aunts also frequently relied on my grandmother for long-term child care. But none of these relatives ever touched cocaine or had any other addictions.

Although Lyndon Johnson's War on Poverty had helped bring the percentage of black families living in poverty down from 55 percent to 34 percent between 1959 and 1969,[2] that progress began to be reversed during my childhood. Unemployment among urban black men rose throughout the 1970s, reaching 20 percent by 1980.[3] The rate for blacks has always been at least double that for whites—and studies find that this bias tends to persist, even when blacks are equally or even more qualified than whites.

And so, atop this clear example of institutionalized racism, job losses driven by industrial contraction and cuts in social services under President Ronald Reagan created vulnerable communities. High unemployment rates were indeed correlated with increases in crack cocaine use, it's true: but what's not well known is that they preceded cocaine use, rather than followed it. While crack cocaine use has been blamed for so many problems, the causal chain involved has been deeply misunderstood.

Indeed, much of what has gone wrong in the way we deal with drugs is related to confusing cause and effect, to blaming drugs for the effects of drug policy, poverty, institutional racism, and many other less immediately obvious factors. One of the most fundamental lessons of science is that a correlation or link between factors does not necessarily mean that one factor is the cause of another. This important principle, sadly, has rarely informed drug policy. In fact, empirical evidence is frequently ignored when drug policy is formulated.

We will see this most clearly when we examine the penalties for crack and powder cocaine and explore the disconnect between spending on law enforcement and prisons and drug use and addiction rates. Crack cocaine, for example, was never used by more than 5 percent of teenagers, the group at highest risk of becoming addicted. Risk for addiction is far greater when drug use is initiated in early adolescence versus adulthood. Daily use of crack—the pattern showing the highest risk for addiction—never affected more than 0.2 percent of high school seniors. A 3,500 percent increase in spending to fight drugs between 1970 and 2011 had no effect on daily use of marijuana, heroin, or any type of cocaine. And while crack has been seen as a largely black problem, whites are actually more likely to use the drug, according to national statistics.[4]

Indeed, when I first learned about actual crack cocaine use rates and the race of most crack cocaine users—among the many other false claims made about the drug—I felt betrayed. I felt like the victim of a colossal fraud, one that had been perpetrated not only against me but also against the entire American people. To understand my story, we need not just to understand the results of one policy but also to explore some of the ways drug strategies have been used for political ends.

As Michelle Alexander brilliantly explains in her magisterial analysis, *The New Jim Crow: Mass Incarceration in the Age of Colorblindness*, American drug policy has often intentionally masked a political agenda. The use of drug policy to "send a message" about race was a key part of Richard Nixon's infamous Republican "southern strategy." That strategy was aimed at winning the South for Republicans by exploiting white fear and hatred of blacks in the aftermath of Democratic support for the civil rights movement. It made words like *crime*, *drugs*, and *urban* code for black in the eyes of many white people. Consequently, it gave legitimacy to policies that appeared to be color-blind on the surface but in reality inevitably resulted in increased black incarceration and disenfranchisement. Even as later administrations continued this so-called war on drugs without necessarily having the same goals, the biased results remained the same.

Indeed, all of the outcomes of these policies—the wasted potential of people behind bars, the shattered families, the missing fathers, the violence seen in the drug trade, even high unemployment rates for black men—were soon being blamed on the nature of crack cocaine itself. I myself agreed with this view in my twenties, even though, as we'll see, my own experience should have made me question it. But in fact, these problems were either worsened or actually created by political choices in economic and criminal justice policy. The policy decisions and misconceptions about the dangers of drugs devastated my generation while we ourselves were blamed for their outcomes. Before I became a scientist, I bought right in.

Meanwhile, the real problems that had made our communities vulnerable to many social ills remained absent from public debate and unaddressed. They are visible in stories like mine, but only if you know where to look and how to think carefully about the problem. It took me many years to understand it. Unfortunately, many people—both blacks and whites—fell for the idea that crack cocaine was *the* key cause of our problems and that more prisons and longer sentences would help solve them.

And now, even though crack cocaine is no longer a major political or media concern, one in three black males born after 2000 will spend time in prison if we don't shift course drastically.[5] My youngest son, Malakai, is in this age group and I am doing my damnedest to protect him by exposing the injustice of this situation.

Of course, children have no understanding of the larger forces that shape their lives—and I certainly didn't know what was going on as the 1970s turned into the 1980s and the maelstrom of economic, political, and criminal justice upheavals of the era began to shred the lives of everyone around me. In fact, I was about to be miseducated on virtually everything about drugs, crime, and the causes of neighborhood strife, including the ongoing domestic violence that would soon shatter my own family.

CHAPTER 2

Before and After

They fuck you up, your mum and dad.
They may not mean to, but they do.
They fill you with the faults they had
And add some extra, just for you.

—PHILIP LARKIN

When my mother returned from the hospital after the fight with my father, she seemed to recover rapidly. We saw her bandages and knew not to say anything. We hoped that was the end of it. But although the hammer fight was not their last one, my parents would separate and divorce not long afterward. Oddly, however, even when I thought that my mother had been murdered by my father, before she came back from the hospital, I don't remember missing her or worrying about her.

Maybe I've just blocked it out because it was too painful; maybe it just came out in other ways. For example, in my family after my parents' split, we gradually stopped calling her "Mom" or "mama." In my teens, we started calling her "MH," an appellation I'd given her after noting the way George Jetson of cartoon fame referred to his boss by using his initials.

Looking back, I think this was a sort of distancing, a wish to deny her the affectionate names others used for their mothers. Because in many ways, for much of my childhood, despite her best efforts, she just wasn't there. After my parents broke up, my mother spent two and a half years in New York, away from all of her children. I now know that she left in search of higher-paying employment so that she could give us a better life. But back then, all I saw was that we were scattered among various relatives.

I'm sure I must have been upset that she was gone but it wasn't something I verbalized. We never knew when she was going away and when she would come back. My sisters now say they felt like orphans. I realize that I did, too. But we didn't share our feelings with each other then. I think I resented my mother for years because I couldn't admit, even to myself, how hurt I'd been.

Already by age six, I had learned to hide my feelings as well as any vulnerability or need. I thought then that this was the only way to protect myself from further hurt, the only way I could properly be the man of the house. I'd begun compartmentalizing. That would turn out to be a critical skill for my emotional survival. To make it work, I wouldn't even show most of my feelings to myself. I'm still struggling with the detrimental "side effects" of this response to my childhood in my relationships today.

I sometimes catch myself thinking that I have revealed too much personal information to someone I care about and start worrying about how it can be used against me at a later date. Often I recognize that my fears are ungrounded, but well-learned behaviors are difficult to change, whether they involve drug use or any other sort of emotional coping tool.

And when I look now at six-year-olds, I can't help seeing how young and vulnerable children are at that age. I realize now that I must have been quietly devastated—but I thought then that I had to be hard. It was the only way I knew how to cope.

*MH and Carl at a family reunion
in the summer of 1978, about six
years after they divorced.*

However, I don't want to blame or judge my parents: they faced severe challenges that I managed to avoid in my own early adulthood. Before either of them had reached the age of twenty-nine, my parents had had eight children. They'd scraped and saved and had bought a nice home together. Their parenting skills were limited by their upbringing and their education. My father, for example, had lost his dad to cancer by the time he was seventeen and had had only limited male guidance throughout his youth. Despite this, both of my parents were extremely hard-working and did what they thought was best for us. For years, my mom worked the graveyard shift as a nurse's aide, doing as much as she could to support her kids. Unfortunately, the jobs for which she was considered didn't typically pay a living wage.

In contrast, when I reached that age I had only one child that I knew about and was on the verge of receiving my PhD: I had resources at my disposal that they couldn't even dream of. It would be easy to say that my parents made poor choices; the reality is that it is impossible to understand their experience and my early life without fully appreciating its context.

And so, putting aside any thoughts of missing my mom, I focused on wanting to be with my father when my parents first split. As a boy, my behavior was continually shaped by my family's notion of masculinity, virtually from birth. For example, when I helped my father mow the lawn or fix the car, I'd get patted on the head or be given other types of encouragement. In behavioral psychology, this process is called reinforcement. The more immediate the reward or reinforcement* following the behavior, the more robust and frequent that behavior becomes in similar situations. And so I quickly learned that emulating my father was what I should do.

In contrast, I was encouraged to play with my sisters when I was very young, but this behavior was no longer reinforced as I got older. It wasn't seen as an appropriately masculine activity for a growing boy. I gradually stopped doing it because this behavior wasn't rewarded. This process is known as extinction. Behavior that was once reinforced but no longer produces praise or reward will eventually be discontinued and that's what happened to my engagement in my sisters' activities.

Similarly, while my sisters would be comforted and soothed by adults if they cried or expressed sadness as young children, my brothers and I were quickly shown by example or experience that displaying such vulnerability was not appropriate male

---

* For simplicity's sake, I use the terms *reward*, *reinforcer*, and *reinforcement* interchangeably throughout the book. To the purist, as a psychologist, I recognize the subtle different meanings of the terms but choose to use them interchangeably in an effort to enhance the readability of the text without misrepresenting the idea being expressed.

behavior. If my sisters were emotionally expressive, that behavior was reinforced. But the boys in my family were actually punished for engaging in such behavior, which decreased the likelihood of us crying in similar situations. Like reinforcement, punishment that has a high probability of occurring immediately after the behavior is more effective. Punishment, of course, is the use of aversive experiences—like reprimands, spanking, or other ways of inflicting pain—to decrease behavior.

I didn't know it then but I was being conditioned by the consequences of my behavior. Through the work of B. F. Skinner and others, I would later learn how those subtle and not-so-subtle reinforcements and punishments profoundly influence our actions. At the time, though, I just knew that what I had to do, what I wanted to do, was become a man. And the best way to do that was to watch and copy my namesake, Carl. I wanted to spend as much time with my father as I could, to get those rewards and avoid being punished, to try to become who I was meant to be. He treated me like I was the center of his world. He taught me how to mow a lawn, how to wash and repair a car, and when I wanted the much-coveted Daisy BB gun, he bought it. With a child's unconditional love, I didn't see any contradictions in idolizing the man who hit my mother and drove her away.

Also, I didn't like some of the alternatives that faced me if my parents split up and I could not stay with my father. My aunt Louise—whom we called Weezy—could not have been happy being saddled with one or more of her sister's children. When we did stay there—and I would ultimately do so for weeks at a time sporadically throughout my childhood—we felt like she would sometimes take her frustration out on us. For example, her children received preferential treatment. If there was a fight or dispute with our cousins, we rarely got the benefit of the doubt. My sister Joyce described feeling like Cinderella when she lived there, with a wicked stepmother and treacherous step-sisters. Even though some of the ways that Weezy treated us were undoubtedly driven by lack of money and being overwhelmed, that isn't something that children can understand. All we saw was that we were not wanted.

Then there was my maternal grandmother's place. At any given time, at least six grandchildren were staying in Grandma-ma's Hollywood, Florida, residence, sleeping on thick blankets on the floor. My mom wasn't the only one of her three full sisters to rely on her mother for long-term child care—but she certainly did it frequently. I've already mentioned that my oldest sister, Jackie, lived with my grandmother. My brother Gary, who was only seventeen months younger than me, also had a permanent home there. He was sent off to Grandmama's even before my parents divorced. Though I was used to sharing my space with a half-dozen or more kids, her house didn't feel like home to me; I didn't feel welcomed. In fact, I was far from her favorite grandchild.

Instead, I experienced some distinct hostility from my maternal grandmother. She was a tough countrywoman who had been raised on a farm in Eutawville, South Carolina. My mother grew up there, too, deep in one of the most rural areas of the South. My grandmother and grandfather had packed up their family and moved to Florida in 1957, just before my mom turned seventeen. That was five years after Willie-Lee, my mother's then-fifteen-year-old brother, was kicked to death by a mule. My grandmother just couldn't take farm life anymore. Still, she'd spent virtually all of her life before that working the fields and facing the prejudice from both whites and blacks that comes from having dark skin, blackened even further by work in the sun. A big woman, five foot eleven and heavy, she kept her long, graying hair in two braids. Her natural skin tone was the same deep brown as mine.

While Grandmama always made sure we had a place to stay, some of my most vivid memories center on her telling me that I was just like my father. Like him, she said, I was ill-mannered, stubborn, selfish, and rude. Like him, she repeated, I'd never amount to anything. Looking back, it's hardly surprising that a mother would see the man who beat her daughter and ultimately abandoned her with eight young children as a bad guy. I couldn't see that, though, as a child. I just felt her rejection of me. Much as I tried to deny it, it hurt.

And what I also sensed was that Grandmama—like most of white America and, sadly, some blacks—seemed to link my father's bad behavior with his blackness. Someone as dark as him could never have been good enough for her daughter, she felt, even though her own skin was dark. Her Mary could do better. Since my skin was black like my father's, that literally colored our relationship.

Much has been written about how racism often makes its victims into perpetrators, how it is impossible to live in a world that hates people with your skin tone and not have this seep into your own dealings with black and white. When I later read Nietzsche's line that "whoever fights monsters should see to it that in the process he does not become a monster," I knew exactly what he meant. Battling twisted prejudices can twist and distort you, often without your awareness of it. Throughout my early childhood, I saw over and over how my grandmother favored the lighter-skinned children: praising them, while punishing or ignoring the dark ones. The conditioning was insidious.

It's not clear to me that she was conscious of this behavior, but it surely reflected the way she had been treated. We were all molded by these attitudes and behaviors before we could even name them. As I'm sure is true for my grandmother as well, I can't even describe my own earliest experiences of racism—it was so pervasive that it's like trying to recall how you learned to speak. You know there was a time before you had language, but it's impossible to remember or to delineate particular incidents or to know what it was like to not know.

Nonetheless, when I sat down with my sister Beverly to research this book, she showed me just how deep it went. In my family, Beverly and I have the darkest skin—and there was nothing subtle about the way the darker children were treated in my grandmother's home. They called us "blackie" or "darkie." Sometimes Beverly was "teased" that way even at home. I would always shrug it off but the tears in Beverly's eyes as she recalled those words made me realize how much it had hurt everyone. Our behavior is shaped over time by sequences and patterns of reinforcers and punishers, often without much conscious aware-ness on our part of how we are being affected. Even racist behav-ior is learned this way.

For most of my early childhood, however, I myself had little direct experience with white people, since I was growing up in a black neighborhood that they rarely visited. But I did see how the children of the people my mom worked for casually called her by her first name—a way we would never be so rude as to address an adult with whom we had not negotiated such inti-macy. And I also saw how my parents and other adults in the neighborhood responded to their power and how cautious and cowed they could be in its presence.

One of my worst memories is seeing my mother break down and cry when confronted by an unsympathetic white bureaucrat about our food stamp allowance, when I was nine or ten. We clearly needed the assistance: I could see how bare the cabinets and fridge were. Yet this woman acted as though my mother were trying to steal money from her personally. At home, MH was tough. She often stood up to my father, who was much bigger and stronger. She never showed much emotion beyond anger about it. But this unyielding bureaucrat's power and petty condescension and my mother's powerlessness in the face of it just broke her.

Indeed, although I don't remember feeling sad about my mother's absence, I'm sure I missed her and was angry that she wasn't around. I was frightened by my parents' fighting, felt powerless over the way I was treated, and was enraged by things like the biases I saw in the world and at my grandmother's house. In my family, one of the few feelings it was okay for males to express was anger—and to do that properly, you needed to have power or else you would be crushed. When I was little, I got crushed a lot: by my mother, aunts, sisters, and cousins. So that was a lesson I learned early as well.

Although I had carefree and childish fun, too, much of my childhood was spent securing status and power in any way I could. If it didn't give you clout or influence, if it didn't make you cool or make you laugh, I wasn't interested. That focus shaped my youth in many complicated and often conflicting ways. As I look back, it's painful because this struggle for respect ultimately marred or even took the lives of many of my peers. I know now that childhood shouldn't be dominated by a preoccupation with status. But to some extent, mine was. This obsession was another key survival strategy that molded me.

So did the stark contrast in my world before and after my parents split. When they were together, the fighting was terrifying, but we lived in a nice neighborhood of young working-class families. It now reminds me of the idealized suburb of TV's *The Wonder Years*, only with black people. The homes were neat, with manicured lawns and flat one-story houses of the psychedelic-faded-to-pastel colors people seem to favor near beaches. Ours was a particularly lurid aqua.

The smell of freshly cut grass brings me back there even now, my dad taking pride in our yard with fruit trees—lemons, limes, oranges, Chinese plums, some belonging to us, others in the neighbors' yards—out back. Our lawn and yard were always extremely well kept, though the chaos of a family with so many young children meant that toys would sometimes be scattered about. My father was especially fond of our lime tree, which grew fruits so large, they looked more like green oranges. He loved to show off those huge limes. Fresh citrus fruits like that remind me of that time before it all changed.

Before the divorce, Christmases and birthdays brought the Big Wheels and Rock 'Em Sock 'Em Robots that we boys coveted; after the divorce, you knew not even to ask for those kinds of presents. Before, our neighbors were mostly intact families, people with decent jobs, adults who believed in the American dream (at least the black version) and had children with similar aspirations. Our neighborhood was relatively safe. We had the occasional break-ins and robberies but no gunfire. Its values were those of the mainstream, that broad swath of mainly white middle-class America that social scientists and politicians use as a measuring stick and try to evoke as a cultural touchstone.

True, one of my uncles had been shot to death while sitting on the toilet in the bathroom of a club, an innocent bystander who was in the wrong place at the wrong time. But that was unusual and it happened far away from our home. That kind of violence didn't haunt our neighborhood. While we didn't live in the Miami of postcard-perfect beaches and Art Deco hotels, our block was clean and tidy. It was occupied by hardworking strivers, the type who sought above all to be respectable.

Afterward, however, although my mom kept us out of the actual projects until 1980 when I was in high school, we moved about once a year and often lived in neighborhoods that were dominated by entrenched poverty and the knot of problems associated with it.

Of course, before, there were also those fights and the fear and the running to the neighbors to call the police. Before, the chaos for us was mainly in our home; after, it was everywhere. And no one bothered to explain it all to us. There was no sitting the children down and telling us, "Mommy and Daddy still love you but we can't live together." My parents weren't much on explanations in general. They lived in a world where you learned by example, not by explanation. You were told what to do, not why, and that was it. You figured it out or you looked like a fool. There wasn't time for childish questions or wondering.

Consequently, when I learned later about research comparing the spare verbal landscape of American childhood poverty to the richer linguistic precincts of the middle class, it really resonated with me. The classic study by Todd Risley and Betty Hart compared the number of words heard by children of professional, working-class, and welfare families, focusing specifically on the way parents talked to their kids.

It was painstaking research: the researchers followed babies in forty-two families from age seven months to three years. The families were drawn from three socioeconomic classes: middle-class professionals, working-class people, and people on welfare. The researchers spent at least thirty-six hours with each family, recording their speech and observing parent-children interactions. They counted the number of words spoken to the child and described the content of the conversations.

The researchers found that families headed by professionals—whether black or white—spent more time encouraging their children, explaining the world to them, and listening to and responding specifically to their questions. For every discouraging word or "No!" there were about five words of praise or encouragement. Verbal interactions were mainly pleasant, enjoyable, or neutral. In the working-class homes, there were also more "atta-boys" than prohibitions, though the ratio was smaller. But in the families on welfare, children heard two "noes" or "don'ts" for every positive expression. Their verbal experience overall was much more punitive.

During my earliest childhood, my family did not receive what was then called Aid to Families with Dependent Children (or welfare as we knew it before President Clinton). But we did do so after the divorce. Moreover, my mother had dropped out of high school in ninth grade. And so her educational background made our home more similar to the welfare group linguistically. MH's relatives—her mother and sisters Dot, Eva, and Louise, who also helped raise us—shared the same disrupted and scanty education. After the divorce, when she returned to Florida, my mother was overwhelmed, with so many children to support. She worked long hours, and so just having the time to do more than discipline us if we got out of hand was almost impossible. My father also faded out of my life as I grew toward adolescence and beyond.

Hence, unlike those growing up in more privileged families, we were brushed back more than we were praised. That may have ultimately helped me to thrive in the critical, skeptical world of science—but at first, it probably didn't do much for my linguistic development.

Even more stunning was the difference Hart and Risley found in the total number of unique words directed at the poor-est children. On average, the professionals' children heard 2,153 different words each hour spoken to them, while the children of welfare parents heard only 616. Before they'd even spent a second in a classroom, the children of professionals had heard 30 million more words than the children on welfare and had many times more positive verbal interactions with adults. Several other studies confirm these findings in terms of the impact of parental education, style of communication with children, and vocabulary on early language learning and readiness for school.[1] Less conspicuous factors like children's exposure to a broad or limited vocabulary and to varying amounts of linguistic encouragement and discouragement can do far more than obvious scapegoats like drugs to influence their futures.

There is little doubt that I was affected early on by my mother's lack of formal education and the limited vocabulary that was used in my home and by most of the people around me. They couldn't teach me what they didn't know. Nonetheless, I did learn many critical skills from them, among them the ability to listen, to patiently observe, and to be aware of myself. I learned to read other people, to pay attention to body language, tone of voice—all types of nonverbal cues. Data from recent studies show that children from working-class backgrounds like mine have greater empathy: they are both better able to read other people's emotions and more likely to respond kindly to them.[2]

As we'll see throughout this book, what look like disadvantages from one perspective may be advantages from another—and ways of knowing and responding may be advantageous and adaptive in one environment and disadvantageous and disruptive in another. Much of my life has been spent trying to negotiate the different reactions and requirements of the world I came from and the one I live in now. Over time, I had to become fluent in several different languages, including the often-nonverbal vernacular of my home and the street, mainstream English, and the highly technical language of neuroscience.

It wasn't long, however, before I began to appreciate what mainstream language could do for me. My awareness of what I was missing rose gradually, from an initial sense that the teachers were almost speaking a foreign tongue when I started school to a flickering awakening to the possibilities that a greater vocabulary and education more generally might offer over time. One incident stands out in my mind. Though most of my primary and secondary educational experiences were dismal, one seventh-grade teacher took an interest in me. She was about twenty-five, with long straight hair, caramel-colored skin, and full lips—one of the few black teachers at Henry D. Perry Middle School and a woman who could get any twelve-year-old boy's attention.

New to teaching, she was on a mission to inspire the black students, to get us to see the importance of academic achievement. Some of the other black teachers tried to protect us by toughening us up and lowering our expectations to reduce what they saw as inevitable future disappointment, but she saw it differently. She taught me the word *sarcastic*, and I remember practicing spelling it and using it at home.

Before that, the only way I'd been able to express the idea of sarcasm was in phrases like "you trying to be funny?" but here was one word that captured a complex, specific idea. Rap music would soon add cool new words like *copacetic* to my life. But it wasn't until I joined the air force and began taking college courses that I fully recognized the power of language.

In my neighborhood, I think our conversations were restricted mainly by our limited vocabulary and inability to pronounce certain words. I remember being embarrassed when I learned from a white high school classmate that the correct pronunciation for the word *whore* was not "ho." Also, I, as well as most of my family, had great difficulties pronouncing words beginning with *str*. For example, I would pronounce the word *straight* as "scrate."

As a result, verbal exchanges in my neighborhood were minimized. Someone might not even reply to a greeting or question, simply looking up and nodding respect with a hint of eye contact or signaling negation with a small, almost imperceptible turn of the head. These signals were all much more subtle than the language. They weren't appreciated or often even recognized at all by mainstream America.

Consequently, my confidence rose when I began to work to expand my vocabulary: I could take charge when I knew more mainstream apt and apposite words. I soon recognized the sheer power that precise language could give me. It was liberating, even exhilarating at times. But as a child, of course, I didn't know what I wasn't being exposed to.

I did learn early on to observe and pay attention before I spoke. Growing up, the worst thing of all was to look foolish or uncool: it was best to stay quiet unless you were absolutely sure you were right. Being strong and silent meant that you never looked stupid. Even if I didn't care much then about being seen as smart by teachers, I certainly cared about not looking dumb, especially in front of friends. Always, I had to be cool.

Another study also captures some key differences between my family of origin and my current family. Sociologist Annette Lareau and her team spent two years studying twelve families, comparing middle-class blacks and whites to poor people of both races. Families were visited twenty times in a month for three hours per visit. The researchers found that middle-class parents—again, both black and white—focused intensely on their children.

In a parenting style that Lareau labeled "concerted cultivation," these families built and scheduled their lives around activities aimed at "enriching" the children's experience—organized sports, music lessons, extracurricular activities linked to school, and so on. Parents constantly spoke to their children and paid attention to their responses, encouraging them to ask questions if they felt anything was unclear or if they were simply curious. Discipline did not involve corporal punishment and was almost exclusively conducted through verbal exchanges: the main idea was to teach moral reasoning, not just obedience.

In fact, children were encouraged to see themselves as worthy of having an opinion in adult conversations and to interact with authorities as though they deserved to be respected as equals (or at least, future equals). They were urged to express their opinions and argue their positions even in disciplinary matters—and these were arguments that they might, by making a particularly strong case, actually win. But their daily life was also highly scheduled and exhausting, at the cost of time spent with relatives or friends.

Life in working-class families like mine was very different. Lareau called their parenting style "the accomplishment of natural growth," and it was based on different assumptions about children. The idea was not to "perfect" children and ensure that their talents were discovered and honed. Rather, children were seen as naturally growing into what they would become, without a constant need for adult direction.

Consequently, children were not the main focus of adult attention. As in my family, children were expected to learn by watching and doing; verbal explanation was not especially important. One of MH's favorite admonishments was "Get out of grown folks' business!" She didn't see herself as a guide introducing us to that world; it was a separate sphere we would figure out how to enter soon enough. So, when we got attention, it was usually for doing something wrong. Then, physical punishments were often meted out.

The use of corporal punishment during my childhood began after the divorce. At that time, we were disciplined harshly and with little chance for appeal or excuses—that was "back talk" or being "hardheaded," not moral reasoning. And it could make it all worse if you tried it while you were on the receiving end of a beating. We got whipped with belts, tree branches, and the cord on the iron. This was a common occurrence until I was about fourteen and started threatening to hit my mother back. But long before that point, it was made clear that in our world, obedience was what mattered and was valued.

Children where I grew up and in Lareau's working-class sample spent most of their time outside of school in unstructured activity, usually playing with cousins and siblings outdoors. Older children were expected to care for younger ones. And adults and other authorities were seen as sources of power, to be respected and feared, not confronted. If we were going to disobey, we learned rapidly to cover our tracks.

Both of these parenting styles have their advantages, Lareau found (although I should note that she did not look at families that used corporal punishment as severe as in my family after my parents split). The middle-class way was not, as some might expect, superior all around. The working-class children were often happier and better behaved. They were much closer to their extended families and were full of energy. They mostly did as they were told. They knew how to entertain themselves and were rarely bored. They were more adept at relationships.

The middle-class youth, however, were much more prepared for school and far better situated to deal with adult authorities. They could speak up for themselves and use well-crafted arguments to come to conclusions more skillfully. This elaborated way of thinking also helped them better make plans that required multiple steps. Essentially, they were more prepared for success in the American mainstream than the working-class children were. And this was true regardless of whether they were black or white. Through this parenting style, middle-class children were being trained to lead, whether intentionally or not.

Meanwhile, the poor and working class were being trained for life on the bottom. Middle-class children were constantly being taught explicitly to advocate for themselves with authorities, while the lower classes were taught to submit without question. Or, if they were going to resist, the poor learned by experience to do so covertly, not openly.

Indeed, covert resistance permeated my early life so thoroughly that it was as natural as breathing. Even today, I feel uneasy and disconnected when I have to do something like pay an outrageously overpriced bill for cable TV or parking. Part of me still thinks that paying full price is for those who don't have a friend who can cut them a special deal. It has taken me several years to begrudgingly accept the fact that I am indeed out of touch with the part of life that was once defined by getting the inside deals.

The idea behind the "accomplishment of natural growth" strategy clarifies a great deal for me about how my family saw its children and what my mother thought her role should be. While MH was obviously troubled and stressed by the overwhelming task before her, she saw her job as mainly keeping us safe, fed, clothed, housed, and out of trouble. Beyond that, she would teach the discipline of hard work and forcefully, often intrusively impose morals and manners. Life was hard and she didn't think it would make it any easier for us children if she coddled us.

Above all, she wanted us to be scrupulously clean, polite, and well behaved. That would make us respectable—we'd be even better than the ill-mannered white children we often saw when she worked as a house cleaner—no matter how much or how little we had.

But as a child, I was infuriated by this emphasis on manners, appearances, and respect for adults. I didn't understand why adults were supposed to be automatically accorded respect, while children could be arbitrarily dismissed and belittled. It didn't seem fair that a child couldn't speak up and be heard if something was wrong, while any pronouncement or action of an adult, no matter how cruel or foolish, had to be accepted unquestioningly. I didn't understand the way the desire for respectability and some semblance of power and control amid poverty shaped the behavior of adults.

Moreover, the emphasis on obedience until you'd reached adulthood didn't always enhance parenting skills. At least for some members of my family, becoming an adult just seemed to mean a shift from having to take often-irrational orders to being able to give them. While my own kids challenge me far more than I did my parents, I value that because I know damn well that adults aren't always right. Of course, I also want them to question and interrogate the world, not to take things on faith without thinking.

And so, while there are many ways in which my parents were certainly neglectful, there are others in which our upbringing provided significant advantages. For one, I learned to be independent and to take care of myself very early in life. Second, I learned to take responsibility—both for myself and for my younger brother, whom I often essentially parented. Finally, my close ties to my cousins and siblings were another important result, though this was another influence that would have both positive and negative effects on my ability to enter the mainstream.

Nonetheless, in my earliest childhood, there was no pleasure I could see in many of the mainstream words—and no power or clout associated with doing well in school. The drive for status was part of what put me at great risk in my neighborhood, while simultaneously playing an even larger role in helping me to escape it.

My mom liked to listen to Al Green on Sunday mornings, his rapturous voice with its sacred yet really erotically charged falsetto high notes filling the house, the record spinning at 33 rpm on our giant console. With bright gospel harmonies and swirling organ lines, mellow songs of love and heartbreak like "Love and Happiness" and "Let's Stay Together" flooded the house: " . . . we oughta stay together. Loving you whether, whether times are good or bad, happy or sad . . ." It was our music, the kind that didn't get played on FM radio, so it was especially esteemed and comforting.

One Sunday when I was seven, however, Mom picked up an extension phone and heard my father talking to a woman who, it soon became clear, was his lover. Most of their fights had to do with real or imagined infidelity. It was a volatile, unstable relationship. And so, driven by rage, MH went coldly and deliberately into the kitchen. She turned on the stove and began boiling a pot full of maple syrup and water. Revenge would be served hot.

Soon my father got off the phone. He was lying in bed, wearing only underwear. Without saying a word, my mother walked into the bedroom and threw the sticky mixture at him, hoping the boiling syrup would cling to his skin. Her anger had taken over. Fortunately, most of the sweet-smelling but dangerous goo missed him. My father did get somewhat burned on one leg, but the vast majority of the sticky mess wound up on the walls or the floor. But now he was enraged.

Terrified, my mom ran out of the house—my dad chasing her, still wearing only his Fruit of the Loom underpants. Typically

*MH in New York shortly after
she and Carl separated
in 1972.*

when my parents fought there was a predictable escalation from raised voices to violence. This time there was no preface. I just kept clear.

And fortunately for my mother, my father did not manage to catch her. It had been raining heavily, one of those intense subtropical downpours, slicking everything outside. Hot in pursuit, my father slipped on the concrete or wet grass, giving her precious seconds to make her getaway. To this day, my sisters believe he would have killed her if he'd caught her. But she had, for once, planned ahead. MH had called her cousin Bob and asked him to pick her up. He was outside waiting in his car. She jumped in. They sped away before my father could catch her. Recovering himself, my father made my sisters clean the syrup off the walls and floor. But that incident did definitively end my parents' marriage.

Everyone went in separate directions at first. My siblings and I were split up living with various grandmothers and aunts. MH went to New York. My father stayed in our house, and after I'd spent just one night with Grandmama, he brought me there to live with him.

I was so glad to be going home. He didn't take any of my sisters or my little brother, just me, his namesake, who was born on his own birthday. That felt right. I was his first son. I was the oldest boy. I was going to have to be the man of the house soon enough. And I wasn't scared of him; I never felt like the violence between him and my mother had anything to do with me.

Carl never hit me: when he disciplined me, it was with a stern lecture or by grounding. My mother and aunts were the ones who got physical with us children. Also, at the time, I saw both of my parents participating equally in their fights. Like any other boy, I admired my father and worshipped him with that blind, childish love that admits no flaws or contradictions. Where I grew up, though, unpredictable events often led to major life changes.

CHAPTER 3

Big Mama

If you're going to play the game properly, you'd better know every rule.

—BARBARA JORDAN

I was about midway through second grade when I started living with Big Mama, whose home was not far from where I'd lived before my parents divorced. I'd lived with my father for a few weeks after my parents separated. Even though I tried my best to be unobtrusive and well behaved because I so wanted to stay with him, he soon found he was unable to care for a young child adequately. My mother also wanted him to sell the house so that she could have her half of the equity. I'd have to live with his mother.

Although we called her Big Mama, she was actually quite short, around five foot two, but broad and big-boned. A proud Bahamian woman who had come to the United States as a young adult, Big Mama wore long, colorful dresses and oversize cat-eye glasses. Though she always kept her hair back in a neat bun, I never saw her straighten it or use any kind of relaxer or color. Her hair was black, only lightly streaked with gray. I loved Big Mama and she stood up for me, stressing first and foremost self-sufficiency and schooling. A black man without an education don't stand a chance, she would always say.

The debate between the philosophies typically associated with Booker T. Washington and W. E. B. Du Bois was represented in my own family in the differences between my paternal and maternal grandmothers. Big Mama was with Du Bois: education was primarily what would advance the race, and staying in school and doing well there was what mattered most. She was grounded in that idea during her childhood in the Bahamas, where education could clearly lift at least some black people into the elite.

In contrast, Grandmama and my own mother thought that getting a trade was more important. Coming from a farming family in South Carolina, they put more emphasis on hard work as a path to success, like Washington did. My maternal grandmother, mother, and aunts on that side of the family all thought that being economically independent first and foremost was more important than book learning—and that was what they saw elevating black people economically, to the extent that was possible in the segregated South. They emphasized hard, manual labor with an immediate payoff, rather than intellectual work, which might never pay off in that punishing and unpredictable environment.

Of course, context was an important consideration for both Du Bois and Washington: both recognized that neither strategy could be pursued exclusively and that in some settings there were limits on what could be achieved through education or business success alone. My grandmothers reflected this complexity as well.

Although Big Mama put more stress on education, she did not see it successfully lift her family in America during my childhood and she recognized its limits in places where racism radically constricted opportunities. Grandmama, of course, had seen that all her life, which is why she thought striving for maximum economic independence was more productive than wasting too much time on school performance.

I would ultimately side with Du Bois on the primacy of education for myself. However, it would be a long time before that became evident, before I even knew that this was a complicated fault line in black history that had intellectual heroes on both sides. And I think much of the credit for my success today belongs to Big Mama and the important role she played in raising me.

Big Mama took a special interest in me and in my second-oldest sister, Brenda. She took me in when my parents split—but Brenda had lived with her since she was a toddler. At that time, my mom couldn't handle raising so many young children so close together in age. Beverly was born just ten months after Brenda, leaving MH with a two-and-a-half-year-old, a ten-month-old, and a newborn. What began as a temporary arrangement after Beverly's birth in April 1962 wound up becoming permanent for Brenda.

I should note here that these kinds of informal child custody transfers were common among my extended family and friends when I was growing up. Many of my cousins and friends lived not with their mothers, but with their grandmothers or aunts. Although the practice of aunts or grandmothers raising their relatives' children has been attributed to the effects of crack cocaine on mothers, again, the rise of these arrangements preceded the marketing of that drug and is much more complicated.

In my family, I'd say that mistrust in or misuse of contraception played a much greater role. My mother, for example, wouldn't take the Pill, because she said she didn't know what was in it. She felt it might sterilize her permanently or could be part of some conspiracy to destroy the black family. We'd all heard about the Tuskegee syphilis experiments and how black men had been left to suffer a curable disease just to allow white scientists to see how it progressively destroyed their bodies and brains.

If we didn't know the exact details—or, indeed, had many of them wrong—there was nonetheless a horrific and genuine basis for our fear. This was always in the background of our interactions with medicine and science. Although we hadn't heard about Henrietta Lacks, a black cancer patient whose cells were used by white doctors without her permission to create a multimillion-dollar biotechnology industry, that story was playing itself out as I grew up. Lacks's cells allowed many important advances—but none of them helped the family whose genes they exploited, who remained poor and unable to afford basic necessities like health insurance. This story was only recently brought to light by Rebecca Skloot in her book, *The Immortal Life of Henrietta Lacks*.

Although there were legitimate reasons for my mother to be suspicious of the white medical establishment, her suspicion in this case may have made life more difficult for her. Since she naturally continued to be sexually active with her husband, she had a child nearly once a year between 1961 and 1969. It was not just my mother alone but also her mother, sisters, and children who had to live with the consequences.

In Brenda's case, this probably worked to her advantage. Perhaps because Big Mama basically saw Brenda as a motherless child, she coddled her. She always tried to make the granddaughter whom she raised feel special and wanted. Consequently, Big Mama supported Brenda's interest in athletics at school, as well as her academic achievement. Brenda was on the drill team and in the marching band; she loved to strut her stuff. Surrounded by white do-gooders who expected her to go to college—and prodded by Big Mama as well—Brenda soon imagined and reached for the same future for herself.

Indeed, Brenda became the most academically serious of my sisters. She would later be the only one of the girls to graduate from college, with an associate's degree in general education from Miami-Dade Junior College. She was the only one of my sisters who didn't have a child in her teens or out of wedlock. She went on to a long and successful career in reservations at Delta Air Lines. To me, Brenda echoed Big Mama's pronouncements about the importance of finishing my education and amplified them. My other sisters and my brothers didn't get this kind of encouragement from adults. Brenda and I also learned lots of practical things from Big Mama, like how to cook and how to take the bus to get around town.

Our grandmother also tried making us take piano lessons. That never stuck because we didn't really practice. The only use the piano in the living room got was when Big Mama played hymns herself or played and sang with Brother Curtis. He and Big Mama were treasurers in the church where she played the organ. I'm not sure if they were seeing each other romantically or not, but he would often come around to play music and to discuss church business. The Bahamian side of my family were Seventh-Day Adventists who went to church every Saturday.

Even though Big Mama disapproved, I tried to avoid church and related activities as much as possible. It was always either boring or frightening: when I believed in God as a child, I saw Him as an angry, unforgiving God who knew I was up to no good and had no tolerance or understanding of my circumstances. He didn't seem to do much for those who prayed. And when the contrast between people's behavior in church on weekends and during the rest of the week became obvious to me—and as my childhood kept showing me just how unfair life really was—I pretty much stopped believing or at least stopped thinking much about it. Later, in my teens, I sometimes even used the idea of God to convince friends to shoplift with me, saying that He would understand us taking from those who have more. But Big Mama's deep and genuine faith sustained her.

She also looked out for me and stood up for me with my father in a way that no one else did. After I'd moved to Big Mama's, Carl was supposed to do the weekend-dad thing with regular visits. Every Friday night, I'd sit expectantly by the front window, watching for his green 1972 Gran Torino. I'd eagerly count down the hours until he was due to arrive. But, sometimes, he didn't come. Or, if he did show up, it would often be late on Saturday rather than Friday evening and he'd be drunk. On at least one occasion, he was so intoxicated when he took me to his place that we had to pull over on the side of the road because he was hallucinating and knew it wasn't safe to drive. We just sat there until it passed.

I didn't mind when he was drunk. I just wanted to see him, even if all I'd get to do was hang out at his house while he slept it off. When he showed up, his drinking didn't make him abusive or unkind toward me. I never attributed any particular effects to it at all. However, I distinctly remember Big Mama getting on his case more than once, describing how I sat and waited so hopefully when he was late or didn't show, and telling him it was disgraceful to treat a child like that by setting me up for such disappointment. It was unusual to see an adult take my side. It stuck with me.

But while Big Mama was smart and strong-willed, she also had some strange ways about her. Like Grandmama, she played favorites. She was intensely loving toward Brenda and me. However, she barely spoke to our other siblings. Indeed, she simply ignored them. In the same way that I reminded Grandmama of our dad, I think my sisters other than Brenda reminded Big Mama of our mom. And that wasn't good: just as Grandmama saw Carl as abusive and not good enough for her daughter, Big Mama saw MH as irresponsible and unfaithful to her son.

Consequently, she was cold, even indifferent to my other sisters. When they came around, like all the other kids I knew they would say hello to the adults as they walked in. This was a non-negotiable sign of respect. But sometimes Big Mama wouldn't even look up, let alone respond kindly and welcome them. The only reason they wound up going to see her at all was that, later in their teens, they wanted to stay out late and not catch hell from MH. They knew all too well that Big Mama wouldn't keep track of their comings and goings.

Big Mama also kept an unusual home. She owned one of the largest houses in Carver Ranches, a black neighborhood in Hollywood, Florida, just north of Miami. The sprawling, three-thousand-square-foot residence had at least six bedrooms. Her husband, my grandfather Gus, had built it for her. It was, in fact, one of the first houses to be built in that community. However, rather than provoking envy, as such a fine, spacious home might otherwise have done, instead her house inspired fear.

Big Mama's place was known as the hood's "haunted house." It got much of its creepy reputation because essentially, no one had done any maintenance on it—internal or external—since Grandpa Gus died of a brain tumor in 1958. Family stories had it that he'd died slowly and painfully and something in his wife was lost along with him when he finally passed.

When I moved in, although she had three of her adult children living with her—Ben, Norman, and Millicent—only rarely did anyone lift a hand to clean the house or maintain the yard. Ben had an excuse: he was slow and may not have known what to do.

Outside, the lawn was brown and dead. In Florida, the sun burns through and destroys anything you don't diligently tend.

On one side, the yard was much bigger than the front lawn, which added to the house's eerie, off-kilter look. Right in the center of that side yard was a massive sapodilla tree, untrimmed and wild. (It grew large brown fruits that were fuzzy like peaches but tasted like sweet cinnamon pears.)

Inside the house wasn't much better. It was infested with scorpions, spiders, and rodents—so much so that no matter how badly you had to go to the bathroom in the middle of the night, you'd hold it because you never knew what kind of scary creature you might encounter. To make matters worse, between the bedroom where I slept and the bathroom was a long, dark corridor. That hallway was definitely a place that you didn't want to explore at night. After dusk, creepy critters seemed to be everywhere.

My cousin Louie, who was about a year older than me, lived with Big Mama, too. He was there because he didn't get along with his stepfather. We both shared a room with twin beds with my grandmother. She'd sleep on one narrow bed; the two of us cousins slept together on the other. Big Mama's adult children occupied the other bedrooms, while Brenda slept in the front bedroom where my grandfather had died. Since his death, Big Mama had never been able to sleep there again.

At night, Big Mama fell asleep to some kind of talk radio, which she kept at high volume. Louis and I would just lie there in that overheated room with her, eventually crashing from sheer exhaustion. But the radio's messages crept in: what we heard over and over was a parade of white guys forecasting doom, predicting complete catastrophe. There was always some world-threatening political, economic, or environmental crisis going on.

At the time, much of the news centered around the horrors of Vietnam, the Watergate crisis at the White House, and the Arab oil embargo. It scared me at first. I became anxious about the stuff they were predicting, fearing overwhelming disaster of one sort or another. I wondered how we would survive. Soon, however, I got desensitized. I realized that nothing was really changing, that the supposedly imminent apocalypse never really materialized. Our neighborhood was in a process of slow decline, but we weren't exactly getting nuked or overrun by communists. I began to tune those kinds of thoughts out. Oddly enough, this forced immersion in bad news and doom-mongering ultimately made me somewhat optimistic, as well as boosting my skeptical thinking.

Louie was also a good influence in many ways. He was a genius at math: the only kid in the neighborhood that I knew who was in advanced classes. I didn't like it when other kids knew more than I did or were better at something than I was, so I kept an eye on what he was studying and even asked him questions about math from time to time. I'd check out the covers of his textbooks, get the names of the teachers he liked. I wanted to be prepared.

Everything around me seemed to reward competition and competitiveness—from organized sports to the games we played on the streets, even board games. From top to bottom, I saw a culture of competition, not only at school and in terms of work but also even in romantic relationships and between family members. Winning matters; nothing is worse than being a loser. I got this message virtually everywhere. It dominated both the mores of the mainstream and of the hood.

Consequently, I wanted to ensure I was a winner in every way that seemed accessible. For example, though I almost always played on losing sports teams, I was also clearly the star of my team—so those losses didn't bother me as much. In math, I wanted to be ready to learn what Louie had learned when I got to his classes the following year, because I wanted to be at least as good as he was. If there was a way that I could win—or even just show that I was capable of winning—I wanted to find it.

A skinny kid who was short like I was, Louie didn't excel at football or basketball, which were the sports I preferred, but he could play baseball. He was a pitcher and was pretty good, too, so long as he wore his glasses. His coach would make him put them on; otherwise he didn't like to wear them. He didn't want to be seen as a geek. But his aversion to geekiness didn't have the roots you might expect. Kids like us didn't automatically opt out of competing for academic excellence, even though it may have seemed that way from the outside.

Where I grew up, nerds, dorks, and other kids who had a reputation for being "smart" in school did not automatically become targets for bullies for "acting white," as the stereotype of poor black neighborhoods portrays it. We didn't scorn nerds any more—or less—than white kids do. We definitely didn't scapegoat them for the reasons that some "experts" have invoked to try to explain some of the persisting racial achievement gap in school. We were no more anti-intellectual than the rest of America.

It wasn't school achievement itself that we saw as "acting white." It's something much more subtle than that. And understanding this complexity is important to understanding my story and to recognizing what's really going on in poor neighborhoods. What was being reinforced and what was being punished was not about education.

Sure, there were some black children who were bullied for "acting white" in the neighborhoods where I grew up. And, indeed, some of those kids were high achievers in school. Some, however, were not. It wasn't scholarly success itself that made people targets. We didn't disdain academic achievement per se and we didn't look down on those who got good grades because of their marks. "Acting white" was a whole different ball game, something that frequently correlated with school performance but wasn't defined by it.

What really got kids labeled as dorks or sellouts and picked on about their schoolwork were their attitudes toward other black people. It was the way they used language to demonstrate what they believed was their moral and social superiority. The kids who were targeted wouldn't speak in the street vernacular that the rest of us used, even on the street or in other informal settings. They wouldn't really deign to talk to us at all if they could avoid it. Their noses in the air, they looked down on us. It was snobbery, not schoolwork, that was "white" to us.

The dorks and L7s (picture it in a kid's handwriting: it means squares) couldn't see any value in things that were important to us, viewing us as ghetto, just like white people did. That's what "acting white" really meant. Kids like this failed to recognize that sports were, for us, often the only way to show mastery. They couldn't see that leadership—even if you were leading the "bad kids"—mattered. They didn't respect loyalty, which we learned to place above all else.

All they valued was what mainstream America did. They thought that made them better than us. They sided with whites in the competition we all felt; they thought that this made them winners and us losers. While some of them might have also idolized sports heroes just as white people did, they certainly didn't want those jocks dating their sisters. A star athlete, as I later became, might have been acceptable when scoring a touchdown on the field or for a quick high-five afterward to show that they knew cool people. But he wasn't someone they'd consider a friend, let alone as a potential romantic partner for the females in their families. That's one of the primary reasons kids who had been labeled a dork or sellout might have been picked on.

By contrast, a kid who did well in school, who showed everyone respect, wouldn't get bullied for "acting white." Instead he'd get support, along with the good-natured ribbing any children—black or white—give someone who stands out in some way. Indeed, the thugs and roughnecks would often try to protect anyone who was doing well, whether in school or in sports, from danger or from problems with the police or other things that might destroy their future.

In fact, it was just this sort of intervention and protection by people—some of whom ultimately wound up in prison, addicted to drugs, or murdered on the street—that saved me more than once and prevented me from doing some really stupid things. It wasn't only athletes who got cheered on for having a path out. We wanted to see everyone that we liked do well, though, of course, as with all humans, there were the usual jealousies and rivalries, too.

But woe betide the kid who thought getting As made him better than you, who didn't give neighborhood kids their proper respect, whether through lack of social skills or true snobbery. That could bring misery. Though some of what we saw as snobbery might have been lack of social skills, we had little tolerance for it. We knew and followed the social code. We needed all the respect we could get. Further disdain from other black people was just too much to stomach.

Our world required exquisite attention to facial expressions and body language, to unwritten rules about status and signs of disrespect. Reading these cues and responding appropriately could sometimes literally mean the difference between life and death. More often, however, it was "only" your whole social life that was on the line. For kids everywhere, matters involving social life feel like life and death, of course. But in the hood, that's even more exaggerated because there are so few other available sources of status, dignity, and respect.

My frequent moves between one relative's house and another's and my constant contact with cousins, siblings, aunts, and uncles helped me to understand quickly the ins and outs of our social code. My desire for status made me pay particularly close attention, sensitizing me to even the slightest signals about who was up and who was down and how that was determined. I observed all of this closely. And these social skills were crucial to my success.

Smart black people tell their children that they have to be twice as good as whites to get half as far. While this is unfortunately still true for academic and business success, I think it's equally if not more applicable to social skills. A white kid might get away with being a socially clueless snobby nerd—but a black child who acted that way would get ridiculed and demolished. Especially among the poor, social skills make a critical contribution to success, one that is often overlooked.

Louie and I both paid heed to these unwritten rules, something that would ultimately cost him a great deal more than it did me. I liked hanging out with him, playing catch and climbing that sapodilla tree in Big Mama's yard. But if our mothers and grandmothers had understood more about what education really meant, we might have also batted around math problems. We would have seen homework as practice—as necessary for school as we knew it was for athletics.

Instead, the adults around us saw school as a quest for a certificate, a stamp of approval you could show around later in life. Rather than valuing the process of education itself and the essential critical thinking skills that can be gained from it, they saw school as a means to an end. Because their opportunities had been limited, because the people they knew who were educated hadn't actually been allowed to move up in management or become anything better paid than a high school teacher or licensed practical nurse, they saw a focus on academic achievement as a distraction, one that would more likely lead to disappointment and bitterness than it would to real success.

They'd never seen academic success genuinely rewarded. And as I eventually learned in behavioral psychology, if you have no experience with a particular reinforcer, it isn't likely to drive your behavior. If you've never tasted chocolate, you're not likely to be especially driven to get some, since you don't even know if you'll like it. Similarly, saying "you gotta get that education," if you have no experience (even vicarious) with its beneficial effects, will not carry much conviction. It certainly won't be anywhere near as compelling as telling your friends about how good chocolate looks after you watched a friend enjoy some— let alone as compelling as if you were extolling its virtues after becoming a connoisseur of high-end chocolate treats.

Consequently, as a result of their lack of experience with true educational success, most of my relatives saw doing anything more than the minimum required in school as a waste of time.

I know I could have been far better at math—a subject that would later be critical in my work as a scientist—if I'd been encouraged in it at home. Math was one of the few subjects that I actually liked. It didn't rely on words I didn't know or terms that could be twisted. It didn't require exposing yourself to correction by the teacher for speaking in vernacular or mispronouncing words the way that reading out loud or being called on in English or history class did.

You could just write the problems out and show how you solved them on the board. Even better, the answers were always clearly right or wrong. I liked that and my teachers soon saw that I was good at it. My math skills were reinforced.

Indeed, my earliest experiences with school were actually pretty positive. Although officials in charge of the Miami-Dade public school system had fought hard for decades to maintain school segregation and our schools were some of the last in the United States to be integrated, busing was finally instituted in 1972, the year I started first grade. My sisters and I were bused.

My school was located in a working-class white neighborhood that didn't look too different from where I lived when my parents were together, with swaying palm trees and well-mowed lawns. And when I started first grade at Sabal Palm Elementary School, there was no obvious resistance to integration. The four or five black kids in my class of twenty-five or so weren't greeted by demonstrators, dogs or fire hoses, or even dirty looks. Nonetheless, some de facto resegregation did begin almost immediately.

Although we started our day with Miss Rose—a young, very nurturing white woman with sandy blond hair, and whom I really liked—for much of the time, all the black boys in my class would be sent to the "portable." This was a small, supposedly temporary outbuilding at the back of the main school. Inside, it looked like a playroom with blocks, trains, and other toys. But most of our time there was spent in small groups, being drilled with flash cards on basic skills like letters and numbers. We were supposedly sent there because we had "learning difficulties."

Soon, though, I was bored out of my mind. Despite the fact that my parents never read to me as a child, I did know my ABCs and 123s. My older sisters had taught me about letters and numbers. I had also been sent to preschool and some kindergarten in a church basement when I was four and five. Because of all that—and because I was an avid watcher of public television's *Sesame Street* and *The Electric Company*—I already knew the alphabet and how to count. But the school assumed that because I was black, I must be behind. So, off to the trailer I went.

One day, however, Miss Rose took me aside and told me that

I didn't have to go with the other black boys anymore. She gave me a choice, saying that if I wanted, I could stay with the rest of the class. Someone had apparently recognized that I actually didn't need extra help. Since all my friends were in the trailer, I was torn. It would not be the last time that I had to make a choice between friends and what might make me successful in school.

And, as I would do repeatedly during my childhood, at first I chose my friends. I happily accompanied them to the trailer, always hoping that this would be the time when we would get to play with those enticing toys. Sadly, it never happened: it was always drill, drill, drill. Soon the boredom proved unbearable. For the first few days, I told Miss Rose I was going to the trailer. When I got out into the hallway, however, I found that I couldn't make myself go. I wasn't going to sit through another second of that deadening repetition, not if I could find a way around it. So, I wandered around the halls, cautious not to get caught.

I discovered that the classroom next door to Miss Rose's was empty. I ducked in there. I stared at the walls. I counted the ceiling tiles. I looked out the window and searched the desks. However, that, too, got old really fast. When I found myself listening to Miss Rose teaching through the wall, I decided that I might as well just stay in class. That's what I did the next day—and kept doing. My grades were all either S for satisfactory or O for outstanding. I didn't get any U's.

My grades would fall over the years, particularly because I refused to do homework. Unfortunately, in my family and in most of the neighborhoods where I grew up, school was seen as a burden to be borne, just like work was for my parents. At home, doing homework wasn't reinforced. Academics and book learning weren't seen as a source of meaning and purpose and future growth. School was just a set of tedious tasks to be endured and got round and through, ideally with the least effort possible. It was an arena for covert resistance.

Today, of course, like other academics, I bring work home because I enjoy the challenge and want to stay ahead of the game—and so do my children. They know they have to do homework to please their parents and do well in school. They get rewarded for doing it and punished for avoiding it. Like I did as a kid, they see school as their job—but for them it's not a meaningless burden, but rather a path to a desirable future.

Of course, they also know that they still face far greater challenges than their white classmates. And they see the downside that comes from bringing too much work home and not being able to truly participate in family life. Nonetheless, they've seen education pay off for their parents and they don't live in a world where all the adults they know who look like them have been thoroughly beaten down by a world that doesn't want them.

Despite all of this, there was a place where black people were allowed—indeed expected—to excel. That was athletics. In my neighborhood, we'd often have impromptu races down the streets or in yards. From early on, I could always outrun all the boys my age and sometimes a few of the older boys, too. Once I started playing organized sports, I most enjoyed football practice. There, for the first time in my life, I felt a real sense of mastery and dominance. I could do virtually all of the drills better than my teammates, especially speed drills. I knew I would be a star, with that cocky certainty that sustains millions of black kids across the United States, facing improbable odds.

Sometimes, not surprisingly, I came across kids who were better than me. But even when I couldn't initially outperform them, I knew I could outwork them. It was written in my name:

I had heart. Moreover, up until junior high, desegregation gave me the odd advantage of being just one of only two or three black kids on my teams. I was virtually always the most driven.

Football was my first love. It is Florida's religion, and has probably never been more so than when I was coming up during the Miami Dolphins' perfect season of 1972. I remember becoming a Dolphins fan the year before that, while listening to the games on the radio with my father. Later, I'd watch them on TV with my brothers, cousins, and uncles. Everyone crowded round the huge color Magnavox as the excitement rose with each victory and the tantalizing prospect of an undefeated run to the Super Bowl came closer and closer to reality.

My idol was Eugene "Mercury" Morris. He was the running back who rushed for a thousand yards that year. He ultimately played in three Super Bowls and was selected for the same number of Pro Bowls. Mercury was quick, fast, and sharp—just the way I wanted to be, like his elemental namesake, quicksilver. Unfortunately, he would ultimately develop a serious cocaine habit, and a 1982 conviction for dealing (later overturned) put him in prison for a mandatory fifteen-year sentence. He served three years.

But for me, watching him was bittersweet long before that ever happened. I could see clearly in his experience how race had an effect on the careers of even the most talented athletes. Although sports are the most meritocratic pursuits I have ever known—sadly, science is still a bit more marred by racism[1]— even someone as profoundly hardworking, talented, and proven as Morris was not unscathed.

For example, it was clear by 1971 that he was Miami's best halfback. He could obviously outplay his teammate, the white Jim Kiick. Nonetheless, it was Kiick who started at halfback that season. Kiick and Larry Csonka, another white guy and Miami's star fullback, were not only teammates but also best friends and roommates. They were known for hanging out together off the field, picking up women. Their drinking and carousing was so notorious that they were soon labeled by sportswriters as "Butch Cassidy and the Sundance Kid" (Kiick was Butch). Not surprisingly, they wanted to continue their on-field partnership the next season, even as Morris's performance clearly showed that he was better for the team.

The rivalry and the obvious racial undertone to the choice of starter was a huge topic of discussion among my male relatives and friends that year. Morris would have led the NFL in rushing yards average per attempt with his 6.8 and 5.5 yards in 1970 and 1971, respectively—but he didn't get enough playing time to attain the needed number of carries to qualify. His performance in training camp was so outstanding, however, that coach Don Shula finally moved him up to sharing the starting halfback position in 1972. That year, he and Csonka became the first two players on the same team to rush for a thousand yards in a season. All of the brothers were cheering him on. His persistence in being the best and its ultimate recognition on the field, where it really mattered, had a huge impact on me.

I knew that I'd never be the biggest guy—but like Mercury, I could aim to be the fastest and the smartest. I might not ever be able to overcome race entirely, but if I worked hard enough, those problems could be minimized. I'd been taught that practice and grit mattered above all, whatever sport you played. That was another lesson that translated into success for me far beyond athletics. I always pushed myself to do more. Unlike genetic factors like height or size, practice was something over which I had total control.

I'd heard NBA Hall of Famer George "the Iceman" Gervin talk about shooting at least five hundred shots a day—that was practice, not some genetic quirk. Larry Bird also mentioned working till he hit one thousand free throws exactly the way he wanted them every day, not stopping until every one of them landed perfectly to return to him at exactly the angle he desired. And Magic Johnson said that when he heard that Bird did a thousand, he'd be sure he did at least two thousand. I could see that the more I practiced, the better I got, and the more time I put in, the better I was on the field when the pressure was on.

Data now confirms that believing in the importance of practice, rather than innate ability, gives people an edge. It turns out, in fact, that some of the praise that parents give their children is not simply benign. When children believe that they were "born smart," they may actually take on fewer intellectual challenges or risks. They become afraid that if they fail, it will prove that they were incorrectly labeled. For example, Stanford psychologist Carol Dweck and her colleagues have shown repeatedly that children praised for natural intelligence perform less well after failure, are less persistent, and choose to take on fewer challenges, compared to those praised for hard work. When they are taught to value practice, however, these differences disappear.[2] I have no doubt that my belief that practice mattered most was a critical part of my success.

Athletics was also one of the few areas where I'd allow myself to fully experience and sometimes even show emotion other than anger. In 1974, I remember actually crying when the Dolphins lost to the Oakland Raiders in a playoff game, which left them unable to defend their title in the Super Bowl. I didn't let anyone know or see that, of course, but even now I can vividly recall every detail of the final play—the so-called Sea of Hands catch. On his way down, after being hit by a Dolphins defender, Raiders quarterback Kenny Stabler tossed the ball toward the end zone and in the direction of Clarence Davis, who caught it for a touchdown in between three Dolphins. Just thinking about it still crushes me, to this day. And every time they lost, which, fortunately for me, was rather infrequent, I would be completely emotionally drained.

Sports were also my real introduction to math. I memorized the statistics of the Dolphins team, figuring out what they meant and playing with them in my head. I learned multiplication by working out increments of 7 for football scoring, 2 for basketball. In the games on the street I wasn't just learning math—I was living it. And it was fun. I only wish my English and history teachers had been able to capture the joy I found in math in football and bring some related type of experience I could connect with into their classrooms.

But though my English teachers usually weren't particularly inspiring, sports did help me to some extent in that subject as well. It was responsible for virtually all of the reading I did outside of school. While I eschewed homework, I'd eagerly consume children's biographies of any sports star I admired. If there was a book about any of the Miami Dolphins, I'd read it and try to apply its lessons to myself. That wasn't reading, as I saw it; that was sports.

Although I'd spent years before that playing on the streets and in yards, I began playing organized football when I was nine. I played in the Optimist League, where I was a standout and often one of just a few black guys on the team. We were called the Driftwood Broncos. I loved it—but there was one thing that I found incredibly stressful. It wasn't on the field. My biggest stress usually came from having to ask my mother for the twenty dollars required to participate. I knew that money was tight and I hated having to push her about it. But while she wouldn't ever say no, she'd put me off, over and over again. I began to dread both being questioned about it by the coach and having to nag her repeatedly.

The Driftwood Broncos eighty-pound football team.
I'm number 22.

This conflict made me feel bad both for her and for myself for having to ask, since we had so little. I resented what seemed to me to be her procrastination; the resulting anger between us was just a tiny illustration of the many, many ways that poverty can put stress on relationships. I sometimes blamed her, even though I knew she was working as hard as she could. Children can't really understand the reasons behind the choices their parents make; they just experience their results. I remember finding this particularly painful. But I'll say this: my mother never interfered with my athletic pursuits, and since sports were the main reason I stayed in school, that made a big difference.

And from the start, even though I was one of the youngest boys on the team, I was the fastest runner. Like Mercury, I played running back and I made a lot of touchdowns. I was proud to wear his number: 22. Few experiences in my life have been better than that moment in the huddle when I knew I was going to be running the ball. That anticipation, that moment of exhilarating possibility—well, it was almost as good as the exultation I felt when I made it into the end zone. I lived for those moments.

CHAPTER 4

Sex Education

Abandon the urge to simplify everything . . . appreciate the fact that life is complex.

—M. SCOTT PECK

I was certain that I'd caught some shameful and disgusting disease; terrified that I'd gotten a girl pregnant. At twelve, I was just beginning to understand the mysteries of sex, just starting to get clued in to what all the drama was really about. It wasn't like I was inexperienced with girls: in fact, the opposite was true. After all, I had five older sisters, so I had plenty of time to observe up close the behavior and desires of the opposite sex. And where I grew up, the girls started chasing you and claiming you even in first and second grade, so each grade from first through seventh, I'd had a "girlfriend." It was these girls who, strange as it may sound, played a crucial role in keeping me *out* of trouble.

Paulette Brown, a long-haired girl who lived a few doors down, was my first-grade crush. We'd peck and hug, but not much more than that. In my neighborhood, the girls set the pace: a cool cat would go with the flow. You didn't want to seem desperate or pushy. A real man made the ladies desperate for him;

he didn't beg or take liberties. That was how the men I looked up to behaved.

And when I was eleven, I distinctly remember walking down the street and overhearing two older girls talking about me. One of them said, "That boy's going to break a lot of hearts someday," and the other smiled and nodded in agreement. That stoked my pride and piqued my interest, of course, but I was too nervous to approach them. I didn't want to undermine my cool image.

By sixth grade, however, I had fooled around with a girl, whom I'll call Vanessa, in a school closet. She was a year older than me; she'd told me to pull my pants down and showed me what she would let me do, all the while keeping an ear out for teachers and lunch ladies. But it wasn't until seventh grade that I fully understood what it was all about.

It was the summer of 1979. Five days a week, I participated in a summer camp program in the park for underprivileged kids, one of many such initiatives that would soon fall prey to Ronald Reagan's budget cuts. They had hired some of the older neighborhood teens to run it, made a few young adults into supervisors, and offered organized sports and activities meant to keep us off the streets. Mostly, it did.

That day, however, I had other plans. A very attractive girl, whom I'll refer to as Monica, had invited me to her house: her mom wasn't going to be home. We'd been talking on the phone and she told me to come by when I went to the park. That summer, everyone was listening to Anita Ward's "Ring My Bell" on their JVC boom boxes. Much to my chagrin, my mother had recently forced me to cut my Afro into a shag (something like a mullet for black people), an injury to my self-image for which I seriously resented her. But I wore some denim shorts, a football jersey, and Chuck Taylors, kicking it seventies-style.

Monica was a brown-skinned athletic beauty. Her breasts

were just starting to bud. Her muscular legs were slightly bowed, which gave her a sexy walk and stance that emphasized her hips. Her brown eyes were a shade lighter than mine. She had a small, delicate nose and wore her hair short, straightened. Monica wasn't on any sports teams—but boy could she run. I'd watched her fly past many boys on the track in PE class. I knew her from school. She lived in a small bungalow on Eighteenth Street near the park. We started on the couch in the living room and then moved into her bedroom.

Soon we were on her bed. We were grinding and kissing, touching all over. We both had our clothes on; it was summer so she must have been wearing cutoff shorts. I was definitely not at all sorry about missing that day's basketball drills. Because suddenly, I had the most amazing feeling ever. It overwhelmed me, was completely beyond my control. Scoring a touchdown, making a clutch jump shot: nothing had ever felt like this. But then there was this sticky stuff in my jeans. For a moment, I freaked out. I had no idea what it was. Still, outwardly, I kept my composure; I didn't want Monica to know about it. It seemed like I had wet my pants. What cool muthafucka wets his pants when he's alone with a girl? I was mortified.

Then I began to imagine all kinds of even more horrible possibilities. Trying to hide my embarrassment, I got up quickly and I'm sure rather abruptly, hoping she hadn't noticed what was now a stain on my jeans. I muttered something about having to get back to my friends in the park. With rising anxiety, I went to find my cousin Anthony, who was sixteen. He would know what to do.

While I kept my eye out for him, the more I thought about it, the more worried I became. By the time I found my older cousin, I was sure that I had some scary and probably incurable VD, which was the term I knew for such problems. Or maybe what I'd done was what got a girl pregnant? I just didn't know.

"Yo, Amp," I said, using Anthony's street name. I started anxiously trying to describe what had happened with Monica. I didn't want to sound like a fool. He let me go on. I looked at him; I suspect my anxiety was visible, despite my efforts. And then with a big smile on his face, Amp proclaimed, "You ain't got no damn disease." Soon he was laughing uncontrollably. "You ain't done shit," he said, then mercifully filled me in on the facts of life.

Because masturbation was seen as less than manly in our circle and because my parents hadn't educated me about puberty and what to expect, I'd had my first orgasm in the company of a girl. I had no preparation for what had happened. My first experiences of pleasure and desire had occurred in total ignorance, completely without expectations or even language. But as soon as I knew what was going on, I was immediately on my way to becoming that heartbreaker those girls had predicted I would be. And although many of them would never know it, my girlfriends played a key role in my success, keeping me out of harm's way and boosting me up when I really needed nurturing.

I had very few conventional people to look up to in my life, male or female, to show me how to have a committed, loving relationship. My parents' breakup and the fighting that led up to it had been driven largely by infidelity: I still don't know what caused that but I'd certainly seen it. Most men I knew had women on the side. I didn't really know anyone who practiced what was preached at church and I didn't know how to navigate this treacherous emotional terrain. Sometimes, I'm not sure that anyone really does.

Although I didn't know about it until I was an adult, my maternal grandfather had long had a mistress whom he spent early evenings with, returning home to his wife by a certain time. He also had children with other women. And several years after my parents split, my mother herself became involved with a married man. I don't mention this to judge my family: if you look closely at virtually any family history, there are complex, tangled relationships and secrets everyone wants to keep hidden.

But in the world I grew up in, people had multiple partners and their relationships were just as often sources of conflict as they were of comfort. For me, my sexual relationships also kept me busy, when hanging out with male friends might well have gotten me involved in much riskier activity. Since my sisters and cousins all stressed the importance of using condoms (although I could have used more direct instruction; the first time I tried one I failed to leave enough room at the end), being with girls was actually a much safer situation.

My mother's boyfriend then was a man named Johnson. Starting when I was about ten or eleven, I worked for his roofing company. Installing and repairing roofs in the relentlessly humid South Florida summer was brutal. But what was almost worse was listening to the guys I worked with talk shit about the boss. They were in their twenties and they'd talk about how if Johnson was in a bad mood, that was because he couldn't decide who he was going to be with that night. They'd go on and on about all the women he was seeing, in what I now know are extremely misogynistic terms. And even though I knew then that my mom was just one of his many options, I couldn't say or do anything about it. It was infuriating.

Consequently, much of what I learned about relationships was learned the same way I'd learned about sex: by watching others, by copying the behavior of the men I looked up to, with very little explicit instruction, discussion, or even thought. From the start, one thing was very clear: you didn't get too attached to women—or if you did catch feelings, you certainly didn't let anyone else know about it.

Sex was a sport and love was a sucker's game. You might profess love if it would get you sex and you might even do things girls wanted you to do that represented commitment to them, like giving them teddy bears or your class ring. But you kept your feelings in check no matter what, and the best way to do that was to always have more than one lady. Cool guys didn't fall in love or limit themselves to just one. They didn't masturbate; they had girls to take care of their sexual needs—and the cooler you were, the more ladies you had. As a star athlete who was soon to become a popular DJ, I was on my way to becoming very, very cool. In fact, my DJ name would soon be Cool Carl.

I lost my virginity for real when I was fourteen. A friend told me that a friend of hers named Kim liked me, and that her mom wasn't going to be home that afternoon, so I should stop by. Kim wasn't really my type, but I thought it might be interesting.

She was clearly experienced. That day, she took the lead. The sex wasn't anything special. But what was really awkward was afterward, when she told everyone in school that we'd done it. That embarrassed me because she wasn't the kind of girl I wanted to be with.

There was a clear but complicated distinction in the hood between nice girls and "hos," one that could devastate girls if they fell afoul of it. Kim, unfortunately, was already heading in the wrong direction. She was already known at age fourteen as the kind of girl you might see secretly but not be seen with. It was okay for guys to sleep with hos, but your rep would suffer if you made one your declared girlfriend rather than a casual "friendgirl." The consequences were much worse, of course, for the girls who got stuck with that label. Most boys—including me—had no idea how it could ruin a girl's life, devastating some girls more than even pregnancies did. Although today I am ashamed about and regret my participation in this cycle, it was the reality I faced as a child.

Marcia Billings, on the other hand, was a good girl—but not *too* good. She was the girl I wanted, with a perfect hourglass figure. She was built and fine. Marcia was about five foot two and weighed around 120 pounds. I first saw her in a McDonald's, after a basketball game when I was fourteen, a few weeks after I'd been with Kim. I awkwardly propositioned her, but she shot me down. All that took was a look and maybe a few harsh words, something like "Keep stepping" or "Nigga, please."

I was shocked; because I was skilled at reading girls' signals, that kind of thing almost never happened to me. But a few months later, my cousin James was dating one of her friends and he reintroduced us. She didn't remember the earlier incident. Now she was quite happy to meet the young DJ who was part of a crew that had begun tearing up the gyms and skating rinks of South Florida. She became my main girlfriend for most of high school. I gave Marcia my class ring and she was the one I took to the senior prom. As much as I was able to at the time, I loved her.

The more time we spent together, the more her warmth and spirit nurtured me. In fact, I was soon spending most of my nights at her house. We watched the Brooke Shields movie *Endless Love* together and I'm sure we saw ourselves in the dangerous passion shared by the young couple in the movie. I knew that she had my back and she occupied much of my time.

My mother was suspicious and even resentful of Marcia at first. She even tried to split us up by calling Marcia a ho and attempting to make me question Marcia's loyalty to me. But when MH finally realized that this was a battle she'd never win—and that she could find out where I was by calling Marcia—she turned around and accepted our relationship. Still, Marcia was

*Marcia and me at my senior
prom in May 1984.*

never the only girl I was seeing. Soon, in an ironic reversal, she would sometimes call MH to try to locate me when I was on the prowl.

In our world, the girls knew the score and they, too, competed openly to win the finest man. It was understood implicitly that the popular guys had other women: it certainly wasn't blindly accepted or preferred and it was often a source of friction, but nonmonogamy was seen as an undeniable reality. Many girls worked it as well as the boys did. This was something that also went unquestioned.

Naomi was another girl I saw in high school—but in this case, one who nearly got me in serious trouble. Light-skinned with a fun-loving but no-nonsense personality, Naomi went by the alias Sweet Red. She was twenty-one but looked and acted much younger. I started seeing her when I was sixteen. One night, we were in the master bedroom of my cousin Betty's house, which she shared with her husband and two children. Betty and Ernest were in the process of getting divorced. Because their conflict over disposal of the marital property meant that no one was home most of the time, my cousin James and I often took girls there. We even had the keys.

However, when Ernest arrived unexpectedly and found Naomi and me in his bed, I had to very quickly demonstrate that Naomi was not his soon-to-be ex-wife and I was not any kind of rival for her attention. He was seeing red, believing that Betty had dared to bring another man into his own home. Fortunately, I was able to calm him down before he pulled his gun, but I was lucky not to be a victim of mistaken identity in my pursuit of Naomi.

Those are just a few of the girls who stick out most in my mind. There were many others, some one-night stands, some longer-term friend-girls. As I mentioned earlier, the mother of my son Tobias was a girl whom I'd gotten with only once.

In terms of sex, then, my adolescence was not one of deprivation. I certainly don't say this to brag. Sexual fidelity and infidelity are a matter of conflict in every society. Here I'd just like to make clear that my relationships with women sustained me emotionally and buoyed me up when I wasn't getting the nurture and encouragement I needed at home.

I will also note parenthetically that my experience shows that you can become a scientist without having been socially inept as a kid. Unlike many of my lab mates, I did not sit at home, fantasizing about unapproachable girls in tight jeans who were oblivious to my existence. I wasn't the science geek alone with my books or the dork who couldn't even talk to a lady. I didn't spend hours with pornography. Indeed, I was so sexually active that some media "experts" might even have called me a "sex addict."

But that wasn't quite what was going on. Instead, my experience illustrates the problems with reducing complex human behavior to simplistic terms like *addiction* and with trying to blame specific brain chemicals for people's actions. Doing so fails to consider the context under which the behavior occurs. It also places an unwarranted emphasis on coming up with a brain explanation, when carefully understanding the behavior and its context could be much more useful in explaining and altering it.

My behavior with girls reflected not just biology but context and experience. It was not just pure sex drive (though that was certainly there) but sex drive modulated by my social context, including family expectations and neighborhood norms. It was about my desire to be cool, the local concepts of cool and how I interpreted them. It was about the rules I internalized—such as the idea that masturbation wasn't manly—as well as the ones that I didn't. It was frankly, also, about a need for comfort and contact. While science must reduce complexity in order to conduct studies, the interpretation of that data cannot simply then be extrapolated back without recognizing these and other relevant caveats.

As a neuroscientist, however, I didn't recognize this at first and I think many of my colleagues still have difficulty doing so. When I started my career, there was great excitement around a neurotransmitter called dopamine, which was believed to explain why people got addicted to drugs. It was even seen as driving behaviors like the propensity for sexual variety. Indeed, some people seemed to believe it could account for all forms of desire and pleasure. And at first, I, too, thought dopamine could answer these kinds of questions. Recognizing why it cannot be the sole answer is an important part of developing a more sophisticated and productive way of understanding how drugs affect behavior and consequently, how to develop better ways to treat addiction.

The green blips on the oscilloscope were coming fast and furious. *Poppoppoppoppop* was the sound accompanying the images, which were generated by the firing of neurons in a region of the rat brain called the nucleus accumbens. I was monitoring the experiment, studying the effects of morphine or nicotine on these brain cells. Previously, I'd operated on the rat, delicately implanting electrodes into the accumbens to measure the way the neurons there would react to the drugs. Although we couldn't tell directly using this technique, we believed we were studying cells that used dopamine as their neurotransmitter, since these were the most common type of cell in that brain area.

It was 1990. I was an eager young college student, working at the University of North Carolina Wilmington. President George H. W. Bush had labeled that year as the start of the decade of the brain. Dopamine was at the center of addiction research. Researchers like Roy Wise and George Koob had propounded the theory that all psychoactive drugs that people enjoy—everything from alcohol to cocaine to heroin—increased the activity of dopamine neurons in this region.[1] This was believed to cause intense pleasure, which in turn produced desire for more.

And, in the case of drug use, that desire was said to be so overwhelming as to "hijack" the brain's "pleasure center," a major part of which is known as the nucleus accumbens. According to the theory, this center was supposed to be activated by "natural" rewards like sex or food, things that would help an animal compete in the evolutionary race for survival. But drugs can increase the activity of dopamine neurons even more than these ordinary pleasures. As a result, with their brains taken hostage by these unnatural experiences, addicts were seen as inevitably doomed to lose control over their behavior. The need to chase more dopamine would leave them begging, borrowing, stealing, dealing, even killing for more drugs as a result. Dopamine was said to make crack cocaine irresistible and crack addicts' behavior uncontrollable.

This "dopamine hypothesis of addiction" had its beginnings in an accidental observation by James Olds and Peter Milner at McGill University, in Montreal, way back in the early 1950s. They had heard in a lecture that a brain network then known as the reticular activating system (RAS) would motivate rats to learn mazes better if it was stimulated electrically. Increasing the activity of the cells in this network appeared to make rats more alert and more successful at remembering the maze. Eager to study this for themselves, Olds and Milner placed electrodes into rat brains (similarly to the way I would later do, though I was measuring activity rather than adding electricity to stimulate the brains of my rats). They tried to place the electrodes so they could stimulate the RAS.

Once these electrodes were implanted and the rats had recovered from the surgery, the researchers placed the animals, one at a time, in a box. Each corner was labeled: A, B, C, D. Whenever the rat wandered over to corner A, the scientists hit a button to electrically stimulate its brain. Most of the rats just wandered aimlessly. But one particular rat would repeatedly return to corner A, especially during the stimulation, as though the stimulation had made this corner very attractive.

Olds and Milner began to wonder if they'd misplaced the electrode in this rat. They decided to examine its brain closely to see where the probe had landed. When the researchers dissected its brain, they found that they had indeed put the electrode in the wrong spot, accidentally landing in a region known as the medial forebrain bundle (MFB).

Initially, the researchers thought they'd discovered that the MFB made rats curious or interested. And that was probably part of what was going on. But in order to try to figure out exactly what was happening, they next deliberately implanted electrodes in this region in other rats. Instead of stimulating their brains manually, however, Olds and Milner put levers in the rats' cages to allow them to stimulate themselves. And, once the scientists let the rodents start pressing the lever, some began hitting it up to seven hundred times an hour.[2]

Though these findings have been exaggerated—in both the scientific literature and popular press—to make it look like no rat could ever "just say no" to this type of self-stimulation, many rats actually didn't learn to self-stimulate and couldn't be trained to do it. As with drug addiction, this is not a phenomenon that can be understood in isolation from the rest of the environment, even in rats. And as with drug addiction, the truly compulsive behavior was seen only under specific conditions.

Nonetheless, Olds and Milner soon realized that they might be on to something much bigger than a way to enhance learning. They'd discovered some kind of joy spot—in fact, the area soon became known as the brain's "reward" or "pleasure" center. Later, in the 1960s, other researchers would discover that the most abundant neurotransmitter in this region was dopamine and that the MFB carried signals between regions we now think are involved in pleasure and desire, such as the nucleus accumbens.

The rats' behavior with the lever appeared to be a model for reward that could be used to study addiction. Now all that was left to do, it seemed, was to figure out how different drugs interact with dopamine and then discover ways to block this. Addiction might be cured, once and for all.

Over time, however, as you've probably guessed by now, it's a lot more complicated than we initially thought. When dopamine's prominent role in reward was proposed, there were only about six known neurotransmitters: dopamine, norepinephrine, serotonin, acetylcholine, glutamate, and GABA. Now there are more than a hundred. Furthermore, we now know that there are specific receptors—or specialized structures that recognize and respond to a particular neurotransmitter—for each neurotransmitter, and most neurotransmitters have more than one type of receptor. For example, dopamine has at least five receptor subtypes—D_1–D_5. We also now know that hormones like oxytocin and testosterone can act as neurotransmitters.

But despite these ever-intensifying complexities, our theory about dopamine's role in reward has not been appreciably revised since it was originally proposed. And, as you will see later, a growing body of evidence casts doubt on this simplistic view of reward.

Nonetheless, when I started studying addiction, I was a true believer in the dopamine hypothesis. I thought that dopamine probably drove sexual and gustatory excess, that it made crack cocaine addicts crazed with cravings. Many of the researchers I worked with were convinced; my heroes were people like Olds and Milner and Wise and Koob, who had made key discoveries through animal research on brain mechanisms involved in reward. I thought that if we could just understand how drugs of abuse interacted with this neurotransmitter, we'd easily develop better treatments—perhaps even a cure—for addiction.

The answers were in this one chemical in this key circuitry of the brain.

Soon, however, certain research findings began to make me skeptical of this idea. These included some of my own. For example, my master's research involved studying how dopamine was removed from the pleasure-linked nucleus accumbens after nicotine was administered. At the time, some researchers were claiming that cocaine and nicotine acted similarly on dopamine in this area, even though data also suggested that rats pressed levers far more times and would work much harder for cocaine than they would to get nicotine.

Indeed, trying to get rats to press levers for nicotine was one of the most difficult experiments I ever tried. I didn't succeed and I'm not alone. Plenty of researchers also failed at this task. (Incidentally, trying to get rats to press for THC, the active ingredient in marijuana, is even more difficult.)

In my master's work, I looked at how nicotine affected dopamine's action in the nucleus accumbens. But what I was seeing was unexpected: nicotine wasn't acting at all like cocaine. Some of the behavioral effects might be similar in some situations, but in this brain region, the two drugs actually had opposite effects.

The oscilloscope that I monitored displayed a line representing how quickly the dopamine activity rose or fell after a drug or saline solution was given. And those lines looked very different when you compared what happened with cocaine to what was seen with nicotine. With nicotine, the line would go up and then fall off more quickly than with saline.[3] But with cocaine, it would go up and stay up much longer than with saline.[4]

This meant that nicotine was increasing the rate at which this brain region "mopped up" dopamine—in other words, nicotine was taking dopamine out of the connection between brain cells (the synapse) where it has its effect, faster than would occur naturally. But cocaine was acting in the opposite way. It was keeping dopamine active in the synapse for longer.

Because this finding contradicted the conventional wisdom and threw a bit of a monkey wrench into the neat story that was being told about dopamine and drugs, there was some resistance to it at first. Charlie Ksir, my PhD preceptor, and I published the first two papers detailing this research in 1995 and 1996. Some researchers did not want to believe that we were correct. Antismoking activists didn't like it, either, because it got in the way of the useful rhetorical claim that cocaine acted similarly to nicotine in the brain, which claim had allowed them to amplify arguments about nicotine addiction by implying that it was just like the nefarious crack.

Soon, however, our findings were replicated and expanded on by other researchers.[5] Years later, in fact, I was approached by tobacco companies, whom I turned down on more than one occasion. They, of course, wanted to enlist me in their efforts to stress the differences between their drug and cocaine. The distinction that we found, however, didn't mean that nicotine wasn't addictive or even that it wasn't ultimately increasing dopamine's activity.

But it was one clue that the dopamine story wasn't as simple as it first appeared. Although both nicotine and cocaine eventually have the effect of increasing dopamine activity in the brain, they do this via quite different mechanisms. Cocaine delays the termination of dopamine's actions, while nicotine causes neurons to release more dopamine in the synapse. Moreover, each drug also has differential actions on a range of other neurotransmitters, all of which actions can result in very different subjective experiences. Smoking cigarettes and smoking cocaine don't feel identical to most people, after all.

And there were further complications. Researchers began to find that dopamine was released not just in pleasant situations;

such releases also occurred during stressful or aversive experiences that were not at all enjoyable. For example, some studies show that dopamine levels rise when animals are stressed by electric shocks or cues that predict painful or negative experiences. Moreover, while animals stop self-administering drugs like cocaine if dopamine is blocked, the same isn't true for heroin.[6] If dopamine were the only brain source of pleasure, heroin administration—indeed, administration of any pleasurable drug—should also cease.

In addition, drugs that release dopamine, such as amphetamine (Adderall), methamphetamine (Desoxyn), and methylphenidate (Ritalin), are used therapeutically, not just on the street. These medications are often prescribed for attention deficit/hyperactivity disorder (ADHD), both in adults and in children. They're also utilized for treating obesity and narcolepsy.

But although there are some cases of abuse, the vast majority of therapeutic users do not become addicted. Indeed, there's some evidence that children given these drugs to treat attention problems are actually at lower risk of addiction later in life than those whose ADHD is not treated with medication.[7] These drugs always cause increased dopamine release: if elevated dopamine-related pleasure alone produces addiction, why don't these patients become addicts, always driven to get more?

The problem is that when we study things like addiction, we're focusing on pathological behavior and ignoring what occurs under the most common, normal conditions. In reality, most drug use doesn't result in addiction. Very little research has been reported about drug users who haven't lost control over their behavior or animals who won't press levers for nicotine or THC. Even less is understood about the activity of the brain's reward system when people engage in the most naturally rewarding behavior of all: sex. We don't know much about how sexual behavior is encoded in the brain and regulated, and it's hard to tell what's wrong with a brain system when you don't know what happens when it works properly.

For me, even in my teenage years, when I was certainly as driven by sex as any adolescent male, it wasn't something that controlled me. I wanted it, for sure, and I was certainly proud of my reputation as a player. But staying in control was paramount. That was far more important to me than any particular girl or experience of sex. In fact, I remember going to basketball practice one day, immediately after having sex with Monica, the girl with whom I'd earlier had that embarrassing first orgasm. I'd been out all night—and was definitely tired when I hit the court. My friend Jimmy Lopez, who was a guard on a rival team, was watching.

"Damn you moving slow; that pussy must've gotten to you," he said. I was horrified by the idea that he might gain confidence and think that he could dominate me on the court. So I never did that again. After that, I'd abstain before games like a boxer: I didn't want to take the chance that sex could make me less agile. I certainly liked sex and spent a great deal of time chasing it but I always stayed in control.

Also, like most of my friends, I wouldn't deign to fight over a girl. We saw that as uncool; it meant that you cared too much. A player didn't act impulsively out of jealousy. He couldn't be seen as dependent on any one woman's love. Of course, you'd step up if someone insulted your lady or disrespected you by flirting with her in front of you, but that was about your own status on the street, not about her. Desire, compulsion, and control had to be more complicated. It just seemed impossible that this one neurotransmitter, dopamine—one found in only about 1 percent of all brain cells—could, by itself, produce uncontrollable behavior when its levels rose and you felt good.

CHAPTER 5

Rap and Rewards

Social support helps to lessen the negative consequences of stress.

—ELIZABETH GOULD

/

The cavernous indoor basketball court at Washington Park Gym was almost unrecognizable at night. The slippery concretelike floor, which I'd cursed as I'd played on it with the City Park team because it was so hard on the knees, almost seemed to thump along with the bass line. The crowd moved in pulsing rhythm, the girls all dressed in their tightest Jordache, Sassoon, or Gloria Vanderbilt jeans, with belly-skimming tops that highlighted their curves. Lights flashed across the packed-in bodies, revealing different scenes and groups as the colors changed. I'd never seen a party like it before—nor had I ever wanted more to be a part of something.

At the center of it all were the DJs, controlling the sound from behind a wooden Formica-covered stand. One of them was dating my sister Brenda. He would eventually become her husband and they are still married to this day. Brenda met Kenneth

Bowe when I was in seventh grade. It was Kenneth, his brothers, and some of my other sisters' boyfriends who would become the closest thing I had in my life to an active father. These men got me into deejaying, at which I soon aspired to shine with the same competitive spirit I brought to athletics. During our weekly dances, they also schooled me on how to be a man.

Brenda had Kenneth take me along to my first dance when I was eleven or twelve. As in much of my social world, the crowd was exclusively black. There were no bleachers at Washington Park Gym, just a regulation basketball court surrounded by open space that could hold several thousand people. When the party started, it seemed like the center of the universe.

I remember the excitement, the scintillating energy, the pounding bass, the sheer joy of being in a crowd merged in music and amped up by surging teenage hormones. That first night, I was tentative because it was all so new to me. In fact, that was one of the only times I ever danced in public, trying not to look like a fool and moving with the crowd. That was before I knew that the cool people were on the DJ platform or behind the booth, just hanging.

Dancing wasn't cool if you had a better way to strut your stuff, like making the scene itself by playing the music or being involved with the guys who did. I felt insecure and unsure initially, but I soon sized up the situation, recognized where everyone ranked in the social hierarchy, and figured out where I wanted to be.

Before I hit high school, I mainly just watched from behind the DJ stand. Observing Kenneth's brother Richard, who was probably the top DJ in all of South Florida then, I learned how to mix and spin, how to work the mic, and the basic mechanics of operating all the sound equipment. We had Technics turntables and QSC amplifiers. JBL and Electro-Voice speakers provided that booming Miami bass. There were enough electronics to fill a room in Kenneth's mother's house, with literally thousands of records squeezed into his shelves.

Soon I could hear what flowed, what kept the party rocking, and how to blend one beat seamlessly into another. From Richard—who went by the DJ name Silky Slim—I learned how to build the crowd up and feed into its growing energy. I could tell what beats were slamming, when to play a slow jam, and how to bring an evening to a climax, to build bumping backbeat onto bumping backbeat until it seemed like the room itself would explode.

Early on, of course, I didn't get much play: the older guys would let me spin a few songs and say a few words, just to see if I could do it. I was still a little kid to them. But when I showed that I was more than a cute novelty, that I was really able to move the crowd, I began to get longer sets, and by the time I was fourteen, I was part of the group itself.

Deejaying at a dance circa 1983.

We were called the Bionic DJs, after the Steve Austin character played by Lee Majors in the hit TV series *The Six Million Dollar Man*. Kenneth had come up with the name, wanting to illustrate the idea that our sound would be thunderous and powerful. Like Steve Austin, we wanted our sound to be amped up, superhuman. Our names were our alter egos, our aspirations.

Mine was Cool Carl. Kenneth, who was about five foot eight and muscular, went by Mr. Magic. He was the serious one in terms of taking responsibility; he arranged venues and coordinated transportation. But in his manner, he was actually a jokester who could do wicked impressions when he let loose. In contrast, his brother Richard was the star performer. Richard was six foot one. He had long eyelashes framing big almond-shaped eyes that made the girls wild. Silky Slim rocked the mic. He was so smooth that all the girls wanted to be with him and all the guys wanted to be him.

Their older brother Cecil—who didn't deejay but along with Kenneth managed the logistics and the money—was known as Dr. Love. He had twinkly hazel brown eyes and a great smile that women loved. Their friend Adolph was called After Death for his initials and he was the fourth man in our group, although he did not emcee. Another Kenneth—a cousin of Kenneth Bowe, in fact, named Kenneth Good—took the nom de rap Captain Good. He did our lighting with strobes, disco balls, and police siren lights. There were also about a half-dozen honorary members, guys who'd be given black Adidas T-shirts with white lettering, identifying them as part of our crew. In exchange for helping us set up and break down the equipment, they got shirts that essentially told the girls that they were "with the band" and carried that kind of currency.

Soon 2,500 people would come out for us on Friday nights, paying two-dollar admission to a gym like Washington Park or a skating rink that we'd rented. I'd take my turn at spinning and emceeing and I'd feel like the man behind the Technics SL-1200 turntables. I knew how to keep the house rocking. I could sweet-talk the girls and have them out of their jeans by the end of the night. I thought I had mad skills.

We kept up with the latest records through a record club; for a few bucks, every week the labels would send us their new releases, hoping to kick-start a hit with club play. Many of them were garbage, but after hours of listening, you'd often come across something that had that sound, something you could build on. At first, virtually all we played was R&B, soul, and funk. In my early days, the big songs were Tom Tom Club's "Genius of Love"; Captain Sky's "Super Sporm"; Herman Kelly's "Dance to the Drummer's Beat"; and Freedom's "Get Up and Dance" (Grandmaster Flash soon sampled that one to death). Kraftwerk's "Trans-Europe Express" also got a lot of play.

When I started going to jams in the late 1970s, hip-hop (or rap, as it was called then) hadn't yet gained much traction outside of New York. Up there, the mother of rap, a then-middle-aged singer and producer named Sylvia Robinson, had founded Sugar Hill Records earlier in the decade. She named it after Harlem's most affluent neighborhood. Robinson was one of the first people to see the commercial potential in the emceeing and beats she was hearing from club DJs and at street parties. She put the Sugar Hill Gang together, choosing guys who looked cool to be the performers—in the same way that male producers had previously chosen sexy women to make up "girl bands."

Sugar Hill's "Rapper's Delight" was the first rap record to win commercial success. Robinson was also behind Grandmaster Flash and the Furious Five, convincing them to record "The Message," which was their big hit and brought a political sensibility to early rap. When I started, old-schoolers like

Grandmaster Flash were just getting their earliest club gigs and innovating by using the turntable itself as a musical instrument, improvising techniques with their hands. Scratching, backspinning, using multiple turntables, mixing musical genres by sampling records—all of this was brand-new back then in the United States, though Jamaican DJs had been experimenting with these tactics for years.

About every other jam we held, there'd be gunfire and we'd all have to duck, but no one ever actually got shot. Wannabe gangsters were just firing their weapons to flex, to show that they couldn't be messed with. In South Florida, our competitors were groups like Ghetto Style DJs, featuring Luke Skyywalker. His real name was Luther Campbell and he's best known now as a member of 2 Live Crew. In the late 1980s, when he became famous, George Lucas sued him for using the *Star Wars* character's name. Also coming up with us were groups and artists like Instrumental Funk, featuring Super Westley J; Opa-Locka DJs with Slick D; International DJs, starring Benjie the Bomber; South Miami DJs, with Tiny Head; and Party Down DJs, with Pretty Tony. Pretty Tony would go on to produce club-banging hits like Debbie Deb's "When I Hear Music."

And Luther Campbell honed the flow that would come out in 2 Live hits like "Me So Horny," during the DJ battles we'd hold about once a month. They'd play on one end of the venue and we'd play on the other. No one really wound up winning, though: both of our groups had large followings that came to hear one crew or the other. Our sound personified what eventually became known as "Miami bass" or "booty bass," which influenced many early hip-hop artists.

Early on, Cecil was the one who really tucked me under his wing. After the dances, everyone would want to celebrate, by cashing in on their stardom and power. When the night had gone especially well, there would be dozens of girls waiting in the wings to see if they could catch the eye of one of the DJs. The older guys usually sent me home at that time because I was so young. They wanted to be alone with the ladies. I knew the rules: if you didn't have the skills or game to get girls out of their clothing, you were a liability and had to go. So at first, I couldn't roll with the older cats when they were on the prowl.

But Cecil took me in, even then. I'd go with him and his groupies to get some food or just back to his house. I was their mascot; their little pet. Watching Cecil, I learned how to talk to girls in ways that were subtle but clearly indicated your intentions.

Although I probably couldn't have described it well at the time, relationships like those I had with Cecil and my brothers-in-law, with my older sisters, girlfriends, and Big Mama, probably protected me from a great deal of harm. Researchers studying resilience to stress repeatedly find that social support is one of the biggest protective factors. And I needed it. My parents had been absent for much of my early life. Even when my mother was physically present, she worked such long hours and had so much else to take care of that I got very little mothering from her.

But with five older sisters—and at least one grandmother who doted on me—I had some strong sources of maternal nurturing, although my sisters were quite young themselves.

People often consider social relationships only as negative forces in drug use. However, what they fail to understand is the complexity of group behavior. Human beings have always devised means of determining who is "us" and who is "them," and the consumption of specific foods or drugs is typically one way of doing so. Teens are especially sensitive to these cues of

belongingness, and so if drug use is the price of group membership, it's one that many are willing to pay.

Some groups, however, mark their boundaries by avoiding
certain types of drug use—for example, athletes rejecting smoking, 1960s hippies rejecting hard liquor in favor of marijuana and
LSD, and blacks avoiding methamphetamine because it is seen as a
white drug. From the level of the clique to the level of the national
culture, behavior related to drugs isn't only about getting high;
it's often used to delineate group membership and social standing.

The social aspects of drug use also change with age. For
example, having children and getting married are both associated with reductions in drug use; one of many studies with similar findings in this literature found that people who are married
are three times more likely to quit using cocaine and those who
have children are more than twice as likely to stop.[1] Similar data
shows that people with close family and romantic relationships
tend to have better outcomes in treatment[2]—and students' feelings of social warmth and connectedness to school and parents
are linked with reductions in drug problems.[3]

The role of social factors is an important part of why the
"dopamine hypothesis" (or any other purely biological explanation) of addiction like those that I first espoused in my early
work falls well short of providing a meaningful explanation of
such problems. It's certainly true that many people initiate drug
use by copying others and that having a social circle that revolves
around drugs can support continued use. But the vast majority
of drug users never become addicted. And, in fact, social support
itself is actually protective against many health problems and
multiple types of risky behavior, including addiction. Indeed,
a great deal of pathological drug use is driven by unmet social
needs, by being alienated and having difficulty connecting with
others.

The majority of people who avoid drug problems, in contrast, tend to have strong social networks. Large extended families like mine, where dozens of cousins, aunts, uncles, and grandparents live close to each other, help prevent the wearing daily stress of living in poverty from being even worse. And these networks can be protective, even when they include drug users. For example, many of the older DJs in our group and their friends smoked weed, but they kept it away from me. My older friends and brothers-in-law wanted to protect me. They weren't moralistic about it. When I was young they felt weed wasn't appropriate for a kid of eleven or twelve and when I got older, they knew I didn't want anything to hinder my athletic performance.

The important role of social connections in pathological drug use was actually seen in the early work on dopamine if you knew where to look for it, and it was also predicted by the behavioral principles originally propounded by B. F. Skinner. Indeed, even in rat models of addiction—which are just models because they cannot reflect all of the complexities of human behavior—it is clear that excessive drug intake is not simply caused by mere exposure to a substance.

This was demonstrated in dramatic fashion by Canadian psychologist Bruce Alexander and his colleagues.[4] These researchers conducted an important series of experiments that have come to be known as Rat Park. Alexander had recognized that the environment in which most lab rats are kept is unnatural for their species. Like people, rats are extremely social animals and get stressed if kept in isolation—but that was the "normal" condition for most rats used in drug research. Thus Alexander wanted to find out whether the lack of rewarding alternatives—what we call alternative reinforcers—like social contact, exercise, and sex would affect their choices about whether to take drugs.

To do so, these researchers created an enriched environment

for the rodents, which more closely modeled their natural habitat. In this enclosure, there were lots of other rats for social contact and mating, interesting places to explore, exercise toys, and dark refuges in which to nest (rats avoid bright, open spaces). Rat Park also offered its residents another amenity: morphine-laced water, sweetened enough to be appealing for rats to drink.

The researchers then compared the morphine use of rats living in Rat Park to that of those kept in ordinary, isolated cages. They found that while the isolated rats quickly took to morphine drinking, the Rat Park rats did not. Indeed, even when the morphine solution was so sweet as to be overwhelmingly attractive to rats, the Rat Park residents still drank much less of it than the solitary animals did. Under some circumstances, the isolated rats would drink twenty times more morphine than their social-living compatriots.

The same kinds of results have now been found with cocaine and amphetamine. For example, rats reared in enriched environments take less cocaine or amphetamine than those raised in isolation.[5]

When natural rewards, such as social and sexual contact and pleasant living conditions—also known as alternative reinforcers—are available to healthy animals, they are typically preferred. There is now a plethora of evidence collected in animals and humans showing that the availability of non-drug alternative reinforcers decreases drug use across a range of conditions.

Indeed, many researchers have found that making sweets available to rats reduces their preference for cocaine and can even prevent them from developing a preference for it in the first place.[6] One typical study in this literature found that 94 percent of rats preferred saccharin-sweetened water to intravenous cocaine.[7] In another series of experiments, in this case with rhesus monkeys, researchers found that the animals' choice to take cocaine is reduced in proportion to the size of the food reward they are offered as an alternative.[8] While people are now using this kind of data to claim that junk food is as addictive as cocaine, this logic is circular: cocaine was supposed to be especially addictive because animals preferred it to food when hungry; now food substituting for cocaine is used as evidence of the reverse.

And contrary to claims that cocaine inevitably leads to child neglect, even in rat models this is not the case. Like human mothers, rats tend to change their lifestyles when they become pregnant and researchers have found that pregnant and nursing rats choose to take far less cocaine than virgin rat females do. While it may not always seem like it, babies are actually powerful sources of reward to their parents.

Similar findings have also been obtained in human laboratory studies that offer cocaine users a choice between the drug and other types of rewards. (One such study, which we did, was described in the prologue.) In another study, cocaine users had the option to snort cocaine under two conditions. In the first one, they had to choose between cocaine and placebo; in the second, their choice was between cocaine and a monetary reward of up to two dollars. Not surprisingly, the volunteers consistently chose cocaine over placebo. However, even though the monetary alternative was small, they chose to take less cocaine when they had that option, compared to when the only alternative they were offered was placebo.[9]

Basically, having choices makes an enormous difference, even when drugs are involved. Cocaine isn't always the most compelling alternative, even for people whose lives seem to revolve around it. It can be extremely pleasant, of course, but at many times, the pleasure isn't actually more desirable than that from sex or other natural rewards. The choice to use depends far more on context and availability of alternatives than we have been led to believe.

Of course, you have probably heard about studies in which rats or even primates continually pressed levers to get cocaine, heroin, or methamphetamine until they died, choosing drugs rather than food and water. But what you probably didn't know is that these animals were kept in isolated, unnatural environments for most of their lives, where they typically became stressed without social contact and had little else to do.

By analogy, if you were in solitary confinement for years with only one movie as a source of entertainment, you'd probably watch it over and over. But that wouldn't necessarily mean that that particular movie is "addictive" or compulsively watchable. You'd probably still watch it if it were the worst film ever made, just to have something to do. Similarly, saying that unlimited access to cocaine "makes" animals addicted to the point of killing themselves, based on research in isolated rodents or primates, doesn't tell us much about drug use in the real world.

Obviously, if you are spending 24-7 alone and without any social contact let alone affection, some drugs, at the right doses, can be quite attractive. However, studying the drug without providing these important alternative reinforcers tells us little about how cocaine affects people or even animals in the natural world.

It presents the drug as uniquely pleasurable and the addicted person as a fool caught in mindless hedonism: it obscures the fact that when people have appealing alternatives, they usually don't choose to take drugs in a self-destructive fashion. But it does show that in the absence of social support or other meaningful rewards, cocaine can be very attractive indeed. The bottom line is that we have been repeatedly told that drugs like crack cocaine are so attractive that users will forgo everything for them. Nonetheless, overwhelming empirical evidence indicates that this is simply not true.

My own social network, however, was also profoundly affected by the stresses of my neighborhood, even as it often helped ease them. Early in my adolescence, one sister, the one I felt most connected to, was nearly taken from me forever. Although Brenda and her husband and his brothers may have had a bigger overall impact on my life, Joyce was the sister I was closest to, both in age and emotionally. She's only a year older than I am. On the outside, Joyce seems tough: we mirror each other in that we both set aside and compartmentalize our emotions. She doesn't take crap from anyone. Joyce is also very sensitive, however, and I think this made our childhood especially challenging for her.

Unlike me and my other sisters, she didn't resist the constant wear of growing up poor and black by trying to stand out or lead. She didn't attempt to be a star athlete like I did or to aspire to college like Brenda. She didn't do well in school the way my other sisters did. She wasn't into high school cheerleading like Beverly and Patricia. She didn't even distinguish herself by surrounding herself with friends with status. In fact, we ultimately grew apart as she began to see me as arrogant. "You think you're better than me," she'd say.

The change in Joyce escalated when MH moved us to the Crystal Lake projects in 1980. These projects, which, ironically, have now become expensive condos, were located in Dania, which is closer to Fort Lauderdale than to Miami. They were two-story brick buildings, built low to the ground. There, for the first time ever, the apartment my mom rented had enough bedrooms that I shared with only one sibling.

But the Crystal Lake projects were zoned to a different high school than the one I'd started at. Since it was Patricia's senior year in 1981, MH didn't try to switch any of us until that fall. Then, however, she wanted us to go locally. I didn't want to make the change. I'd established myself at Miramar and had standing in sports and a tight group of friends. So I stayed true to my school, splitting my time mainly between my girlfriend Marcia's house and Big Mama's, which were nearby, and only occasionally staying at my mother's new, more distant apartment. Joyce, however, agreed to switch schools and began attending South Broward. I started to see her less.

When she got shot in an incident that reverberated through our social world, we were just beginning to drift apart. Joyce wasn't the intended victim: that was Kenneth Good, who would later become the lighting man for our DJ group. I don't even know what the beef was about, but a guy whom I'll call Wes— and who had dated my sister Patricia in junior high—had a problem with Kenneth. Wes was then in high school, maybe sixteen or seventeen, short and stocky. Whatever the issue was, Wes had taken it seriously enough that he'd planned to shoot Kenneth over it. No one knew when it would happen. When trouble was coming, we usually could sense it, but the timing here was a surprise.

We'd all gone to a high school football game. I wasn't playing, but Beverly was cheerleading. Some of my female cousins were there as well. It was sometime in 1979 and I was twelve or thirteen. I'd started deejaying but wasn't getting much play yet.

After football games, everyone always went to a nearby McDonald's in Hollywood. It was across the street from the city's major mall, the Hollywood Fashion Center. Hundreds of people would flood into the large parking lot. Under the palm trees, music was bumping at volumes aimed at displaying the raw power that could be achieved by a carefully selected and modified car sound system. KC and the Sunshine Band's "Do You Wanna Go Party" was one of the biggest hits that year and I'm sure they played it at least once that night. Bright street-lights, almost like floodlights, kept the parking lot lit up.

With such a large crowd, the line for food already stretched almost to the door when I rolled up with my cousin James. Joyce was standing near the doorway, probably next to Beverly and near my brother Gary. A crowd of people was gathered there, including Kenneth, laughing and talking, maybe deciding whether it was worth it to get on line then or wait.

We'd just parked when several shots flew across the parking lot. It was maybe ten thirty or eleven at night but the garish lighting made it pretty easy to see. I was starting to step out of James's car. I heard a sudden, familiar *tat tat tat*. Everyone knew instantly that this wasn't some firecracker or car backfire. We all hit the ground. We knew the drill. It was far from the first time I'd seen gunplay.

In fact, not long before this, I'd seen a white guy get shot and killed outside a park where I sometimes played basketball. He'd been killed in retaliation for the shooting of a sixteen-year-old black teen whose street name was Flap, the older brother of a boy I knew. I'd seen how that death had changed his family. My mother was close to his, even though I didn't know him or his younger brother that well. I'd kept a lid on my feelings about all of it, trying to seem nonchalant as I watched the white guy fall dead to the ground and then learned about what happened to Flap. It was hard to believe that moments like that could end a life.

Of course, when the shooting starts, the thought that you might get hit is inescapable. Everything seems to go into slow motion and your senses heighten to take in every sight and sound. Memories rupture into snapshots. The next thing I heard was Joyce shrieking desperately for my sister Beverly because she, Joyce, was hit. She was on the ground, bleeding and just screaming and screaming. Beverly was holding her.

Wes was hanging out of the window of a car, with the huge black barrel of a shotgun pointing toward the crowd at the McDonald's door. My sisters and brother Gary were still vulnerable. I saw Wes pull the gun in. Then whoever was driving him began pulling away.

Someone called an ambulance, which arrived almost immediately because we were close to Hollywood Memorial Hospital. By the time the EMTs arrived, staff from the McDonald's were already with my sister, bringing out whatever they had on hand to try to stop the bleeding. She'd been shot in the head with buckshot and her face was drenched with blood. I was afraid that she'd die. I thought about how we'd once been so close. But my sadness and concern were quickly replaced by anger and a desire for revenge.

No one talked about those thoughts. Or rather, those who talked about get-back were quickly discovered to be braggarts or cowards who wouldn't actually do anything. We weren't so stupid as to incriminate ourselves like that. You might say a few words like "That motherfucker needs to get his," but it was your body language and rep that spoke for you. It showed you were a man.

What seemed like only a few seconds later, the police showed up with Wes in the back of their car. They asked me to point out the shooter. I looked straight at him. He was desperately trying to seem hard, but I could tell that he was really terrified. He looked diminished and shrunken somehow; in handcuffs he seemed like a child. I pointed my finger accusingly, acknowledging to the police that he was the one I'd seen with the gun. You didn't protect the kid who had shot your sister from the police. But I also wanted him to pay with more than an arrest and conviction.

Meanwhile, my cousin Wendy had gotten into the ambulance with Joyce, holding her hand and trying to console her. Beverly stayed back, trying to reach my mother to let her know what was going on. I didn't know it then but the fact that Joyce had remained conscious suggested that the wound wasn't that bad. It turned out that she'd been hit over her right eye and on her tongue. She escaped being blinded in one eye or worse by only inches. But the doctors were unable to remove the buckshot from her tongue, which remains there to this day.

However, she stayed in the hospital only a few hours that night, until she was stabilized; she returned a few days later to have plastic surgery on the wound over her eye.

All that time, I focused on revenge. I was young, but I knew that men didn't tolerate that kind of offense against their family. If I didn't stand up for my sister, my reputation would fall. It didn't matter that she wasn't the intended target: Joyce was the one who got hurt. But there was a complicating factor: Wes's family and mine had been close. My sister Patricia had previously dated him and I'd dated his sister Lisa in middle school. Our mothers were good friends and whenever I visited, Wes's mom had been especially kind and welcoming to me. I also liked his brother.

Still, while I waited to find out whether Joyce would be all right, I thought about how to get back at Wes. I tried to get a gun, but at twelve or thirteen, I didn't have friends my age who had guns, though many of them pretended that they did. Guys who had real access wouldn't take me seriously. I think they were trying to protect me from doing something stupid. And even if I had managed to acquire a weapon, I didn't know how to find Wes. He'd been taken away immediately to juvenile prison. There wasn't really more that I could do.

By the time I saw him again, everyone had moved on. To the family, Joyce seemed fine. Amazingly, she wasn't even disfigured. Thinking back over the course her life later took, however, I wonder now about how traumatic it must have been for her. She went back to school just a few days after the shooting. This was not the age where people received counseling for possible psychological distress. And once we knew she was physically okay, no one said another word about it in the family.

Joyce was left alone to grapple with having had a profoundly life-threatening experience. No one in the family realized that she needed extra love and support; we all thought that once the physical wounds healed, she'd be fine, and she behaved as though she was. But Joyce would ultimately be involved in a number of violent incidents, two of which stand out. Once, she got stabbed by a woman who was angered that they were both seeing the same man; another time, she stabbed a different woman in a similar dispute.

For most of her twenties and thirties, her life was chaotic and unsettled. But it's interesting to note that despite all this, she never had any kind of drug problem. Her issues were related to her relationships and, possibly, her experience of that trauma. Sadly, she would later blame me for leaving the family to join the air force as she was left dealing with these events, saying that I'd failed as a brother by not being there for her at that time. None of us realized back then that such support was supposed to come from parents and other adults, not siblings who were just children themselves. Her feelings of disappointment still pull at me.

As for Wes, he was incredibly apologetic when he got out

of juvie. He said over and over that it had been an accident. He certainly hadn't meant to hurt Joyce. Our families stayed close, and since Joyce seemed physically fine, we put it behind us. I wouldn't get my hands on a gun until the idea of getting back at Wes for shooting Joyce had long been discarded.

CHAPTER 6

Drugs
and Guns

Only by learning to live in harmony with your contradictions can you keep it all afloat.

—AUDRE LORDE

t was Richard's grandfather's gun, a large rifle that looked like an M16 but shot .22s. It wasn't a handgun that you could hide down your pants, so we usually kept it in the trunk of my car, a 1972 Pontiac LeMans in midnight blue with a white vinyl top and a suave cream leather interior. I'd paid four hundred dollars for it. I was planning on putting Tru-Spoke rims and Vogue tires on it, but never got around to it. I was sixteen, just entering my senior year in high school. I was at the wheel and Richard, whom we called RAP III, for Richard A. Ponte III, was almost literally riding shotgun. He held the gun across his lap as we headed home.

We were driving down Hallandale Beach Boulevard, just coming off I-95. It was a four-lane road that marked the border between Carver Ranches and a white neighborhood. We were probably returning from eating at a local Denny's, a place we frequented with an irregular policy of "dine and dash," sometimes failing to pay the bill. We were bored.

Then I noticed someone walking along the side of the road. That in itself was unusual: this was South Florida, and people drove, they didn't walk. But what was really strange was that it was a white guy.

"What he doing here?" someone said.

In the back of my car were the two Derricks, my good friends Derrick Abel and Derrick Brown. No one ever called Derrick Brown by his given name. Since elementary school, he'd been "Melrose," after the local school for developmentally disabled kids (whom we then called retarded). He wasn't really any more "retarded" than the rest of us, but he'd tested badly in school and the name had stuck. Melrose was slightly taller than me, about five foot ten. His skin was a dark, blue black and he was built. Most of my teenage friends looked immature compared to the well-developed young women around us, but he looked like a man, with a huge chest and arms.

Derrick Abel was something of a mama's boy. His mother was a Jehovah's Witness and she tried to keep close tabs on him. We called him Super Slick, but the name wasn't as resonant as Melrose was for the other Derrick. Sometimes it seemed aspirational or almost ironic. With his strict mother, Super Slick always felt he had something to prove.

Though his mom blamed us for being a bad influence, much of our misbehavior was, in fact, instigated by her son. He was tall and very thin, with the close-cropped hairstyle we all wore at the time. We thought the more flamboyant eighties hairstyles like Jheri curls were uncool. Like the rest of us, Derrick dressed in tightly pressed high-water pants and short-sleeved Izod shirts. He was constantly trying to show how tough he was.

In this case, though, it was probably my idea to mess with the white guy. As usual, Slick joined in and no one dissented. We didn't consider any possible consequences or even think at all about what might happen if things went wrong. We just thought that the guy was out of place. He was on the border of our turf and this particular intrusion by a white man was something that we didn't have to tolerate. We had the power here.

As we came up from behind him, I slowed the car to a crawl. By then, Richard had positioned the gun in a menacing position, rolling down his window and sitting as though he was taking aim. "Put yo' hands up, muthafucka!" he shouted. The dude froze.

I will never forget the complete look of terror on that man's face. His eyes opened wider than I thought it was possible for eyes to go. He was standing still but clearly shaking. His heart must've been pounding out of his chest. He was probably just heading home from work, an ordinary guy in his twenties, wearing jeans and a T-shirt. I'm sure he never expected anything like this. Looking back, I realize it must have been incredibly traumatic.

At the time, though, we thought it was hilarious. The four of us started laughing when we saw the look on his face. I'm sure he thought we wanted to rob and/or kill him. But that was not our intention: we thought we were just messing around. Our laughter must have seemed stone cold. In retrospect, I have a hard time imagining how we could have done it, given the terrible toll we'd all experienced from gun violence. Still, we had nothing particular in mind. It was just an impulse, one that could have had terrible consequences but fortunately didn't. Richard stared at the dude, keeping the gun aimed squarely at him. After a few more seconds, the guy's instincts must have taken over and he ran like hell. We just drove away.

The whole thing couldn't have taken more than a minute, but the image of that man's fear and the sense of power we had—as well as, I see now, our heedlessness—has always stuck with me. I can see the world from other perspectives as an adult, but back then, I really couldn't. My concerns were entirely focused on the respect of my peers and whatever was necessary to maintain my status. I just didn't see that white guy as human; he wasn't one of us. We kept laughing and going over what we thought were the funniest parts of his reaction.

"You saw that muthafucka's face?"

"I bet he nutted on himself."

"Damn . . ."

As I grew up, I maintained a complicated relationship with the street. First and foremost, I saw myself as an athlete. Sports and girls kept me busy at many times when cousins and friends were getting into troubling incidents that didn't end as well as that one did. Sports also gave me the typical "jock" perspective of skepticism about things like smoking that might interfere with performance. First football and then, for most of high school, basketball were the primary reasons I went to school: while I practiced intensively and with great commitment in sports, I did only the bare minimum schoolwork needed to keep up the 2.0 average required to stay on the team.

My expectations about school had always been low, but not as low as most of the educators' expectations were for me, with a few conspicuous exceptions. Here's one example: My senior year, one of my classes was parking patrol. Just as it sounds, we just sat there and watched cars in the parking lot. I'm not sure what it would have taken to fail that class but virtually anything would have required more intelligence than it took to pass it.

Another example involves the end of my engagement with real math in high school. In ninth grade, I'd actually been placed

Shooting a layup during a high school basketball game.

in one of the highest-level math classes. I had continued to do
well in math throughout elementary and middle school, despite
my refusal to do homework. But then I tore up my knee playing
football and had to have surgery. It was after this that I switched
from football to basketball. Before my injury, I'd excelled at algebra. However, because I'd missed so many classes when I was in
the hospital, school officials told me I didn't need to finish out
the semester in the top class. Instead, I could take business math,

which was basically addition and subtraction, third-grade-level stuff. That completed my math requirement—and therefore my math classes, period—for high school.

Rather than challenging me to learn, they gave up, figuring that it didn't matter because I was just one more nameless black kid who would never go to college anyway. And of course, given an easier option and no reason to challenge themselves, almost any teenager—and most adults, too—will take it.

And so other than two to three hours of daily basketball practice—and of course, games—I barely spent any time in school. I'd been put on the "vocational-tech" track, which meant that I received school credits for working as a busboy at the café at Walgreen's. I'd have class from eight to eleven; then I'd go to work. One-third of the time I spent in supposedly educational programming consisted of classes like parking patrol. But I always worked as many hours and as many jobs as I could get, following my parents' example of being hardworking.

Still, none of this meant that I didn't sometimes engage in the same types of petty and ultimately not-so-petty crimes that people so often falsely attribute to the influence of drugs. The incident with the gun was only one of many criminal acts for which, luckily for me, I did not get caught. Starting at about seven, for instance, I'd been tutored in shoplifting by cousins Amp and Mike. Although a large proportion of the people in the neighborhood where I then lived were on welfare and receiving food stamps, no one wanted to be seen using them.

In fact, we mercilessly teased people who were caught showing the multicolored bills when they were sent to the store to get milk or other groceries. There weren't any supermarkets in the neighborhood, so we shopped at a strangely named chain of convenience stores called U'Tote'M, which was bought out by Circle K in 1983. They were usually owned by whites or Middle

Eastern people. They hired white staff, usually bored teenagers who didn't care much about their merchandise or their jobs. That worked in our favor.

When my parents were together, we'd had no need for food stamps. But after they split up, I would be sent to buy groceries with them. It wouldn't take long to find the few items like milk or eggs that were needed. What did take time was making sure I wasn't seen checking out without cash. I'd hang back and wander the aisles, trying to make sure no one I knew was around. When the coast was clear, I'd pay. After my cousins taught me to shoplift, however, I started using what I'd learned taking candy bars and potato chips, to do the household shopping, too. This was another way I showed my cool—and got some much-needed extra loot.

Our techniques weren't exactly sophisticated. We'd wear baggy clothes and have someone distract the cashier while the rest of us tried to surreptitiously slide what they wanted under their shirt or down their pants. If the clerks had cared at all, they probably would have caught us, but I always got away with it. The only time I saw a kid get caught was when my cousin Bip slipped a comic book under his white T-shirt. The bright red of Spider-Man was clearly visible through the fabric. The clerk saw it and opened his mouth to start shouting at him.

Immediately recognizing what had happened, Amp took charge. He began lecturing Bip himself. "I'm going to tell yo momma!" he yelled. "You know that's wrong, what were you thinking?" He went on moralizing, while the clerk glowered, distracted by Amp's speech from calling the police, searching the rest of us, or continuing his own lecture. He had no idea that Amp had put Bip up to it; nor did he know that we had our own stolen items concealed in our clothes. When Amp finished his performance, the clerk just glared at us and said, "Out." Bip was thoroughly embarrassed.

Outside later, we gave him even more hell, not just for getting caught but also for stealing something as useless as a comic
book. Other than my sports books, none of us read anything, so
we thought that stealing something to read, even a comic, was
the height of hilarity. But Bip was so shaken by the whole event
that I don't think he ever stole with us again. He would later, in
his twenties, serve time in prison for cocaine trafficking.

Several other kids in my family also shoplifted from time to
time. One of my sisters had a particular knack for changing the
prices on items to get expensive items for almost nothing. This
was before electronic tagging and inventory systems rendered
her method obsolete. I was much more circumspect in what I'd
do. It really had to be a sure thing for me. I had no intention of
ever getting caught. For instance, when I was in middle school,
we'd often hang out at a mall that was at the transfer point for
the bus home. I never shoplifted there: too many cameras and
security guards.

In my own life, then, it was very clear that crime wasn't
always, or even very often, driven by or even related to drugs.
Most of my peers shoplifted, whether or not they took drugs.
Guns, similarly, had little connection to drug use or dealing in
our lives. For us, shoplifting was not a matter of "stealing to support a habit" and we didn't carry guns to "protect dealing turf."
We stole because we didn't have what we needed or wanted; we
stole to resist, to not be suckers. We kept guns to be cool. It was
much more about necessity and poverty, about power, not just
pleasure.

At the time, I didn't think critically about any of this. And so,
when crack cocaine came along, I completely bought the party
line about its connection to violence and disorder. I had similarly accepted without thinking the idea that drugs like heroin
and even marijuana caused violence. I was soon seeing crack the

way everyone around me did: as a scourge, the source of all our problems. I thought the drug itself was what made our neighborhood into a war zone.

But evidence from research tells a different story. It is true that addiction and crime are correlated. People involved in crimes like burglary, larceny, and robbery are more likely to be addicted to drugs than those who don't commit such crimes, and vice versa. However, around half of all people with drug addictions are employed full-time[1] and many never commit crimes that aren't related to the fact that their preferred drugs are illegal.

The U.S. Justice Department's Bureau of Justice Statistics examined the connections between drugs and crime in prisoners, analyzing data from 1997 to 2004. It found that only a third of state prisoners committed their crimes under the influence of drugs and only around the same proportion were addicted.[2] That means the overwhelming majority were not intoxicated or addicted during their crime—and only 17 percent of prisoners reported committing their crimes to get money to buy drugs. Violent offenders were actually less likely than others to have used drugs in the month prior to incarceration.[3]

The real connection between drugs and violent crime lies in the profits to be made in the drug trade. The stereotype is that crack typically causes crime by turning people into violent predators. But evidence from research shattered this misconception. A key study examined all the homicides in New York City in 1988, a year when 76 percent of arrestees tested positive for cocaine. Nearly two thousand killings were studied.[4]

Nearly half of these homicides were not related to drugs at all. Of the rest, only 2 percent involved addicts killing people while seeking to buy crack cocaine and just 1 percent of murders involved people who had recently used the drug. Keep in mind that this study was conducted in a year when the media was filled with stories warning about "crack-crazed" addicts.

Thirty-nine percent of New York City's murders that year did involve the drug trade, however, and most of these were related to crack selling. But these killings were primarily disputes over sales territories or robberies of dealers by other dealers. In other words, they were as "crack-related" as the shoot-outs between gangsters during Prohibition were "alcohol-related." The idea that crack cocaine turns previously nonviolent users into maniacal murderers is simply not supported by the data. When it comes to drugs, most people have beliefs that have no foundation in evidence.

My own drug use was completely dissociated from my other criminal behavior. I didn't slow my car to let Richard point the gun at that white guy because I was crazy from being high or wanted money to get high. And we didn't keep the gun on hand because of drugs, either. I never shoplifted or sold marijuana because I needed money to smoke it. In fact, I actually didn't like marijuana much. By sixteen, I'd tried cigarettes, reefer, and drinking but, as always, my main goal was staying cool. That meant low to moderate use: I didn't want to feel out of control, ever, and I could see how getting drunk or really high could interfere with this desire.

My priority was athletics. I wasn't going to do anything that might impair my performance on the basketball court. Switching my primary sport from football to basketball in high school because of my injury had already put me at a disadvantage. While I was playing football for hours and hours every day in elementary and middle school, most of my teammates and competitors had already been focused solely on basketball. But back then, I'd played basketball, both organized and in pickup games, only as something to do in the football off-season.

I tried to make up for the years of practice I'd missed by playing relentlessly at night, even when I'd already had a few hours on the court that day at school. Sometimes I was the only guy shooting hoops at 2 a.m. in the projects where my family had finally resigned us to living. No matter what was going on, I always practiced at least two to three hours a day. And then, if I was angry, bored, couldn't sleep, or was just sick of dealing with people and their drama, I'd go out and do even more drills and shots, rarely tiring of ensuring my skills were on point. (I now realize it must have driven the neighbors crazy, given that the court was in the center of the projects in an open plaza surrounded by ten buildings.) The summer between eleventh and twelfth grades, I was on three different teams and must have played in practice and games for at least six hours on most days then, often more.

All those kiddy biographies I'd read about athletes stressed hard work and infinite practice. They said that drugs were bad, that smoking anything could hurt performance. They heavily emphasized believing in one's own inner strength and willpower, reinforcing the American ideal of the self-made man, the guy who triumphs through sheer persistence and unending grit. They showed me that the way to win was to outwork your competitors and use everything you had to maximize your skills.

And so, though everyone else thought my height was a disadvantage—I was five foot seven on a good day—I chose not to see it that way. I was a point guard. I didn't have to get in there and rebound with the trees. My job was to distribute the ball. I was always one of the quickest people on the court and had exceptional ballhandling skills. If I got to the basket on a big guy, well, either I was going to score or he was going to foul

*Shooting a free throw during a
high school basketball game.*

me, I didn't care. I was fearless. I'd take it right to you. It was to
my advantage that the bigger guys often didn't expect that, but
I wasn't going to let anyone punk me. I came from a neighborhood where at any time, you might have to fight to defend your

reputation, facing violence that might turn deadly. I brought that level of intensity to the court. The worst thing you could do was foul me. Okay, I get two free throws. I could handle that.

By eleventh grade, I'd moved up from junior varsity to varsity. By senior year, I'd be most valuable player on a team that, because we had a seven-foot center, was thought to have a good chance at the state tournament. My junior year, though, for the first time ever in my life, I just rode the bench. That was because I'd switched sports and wasn't up to the level of the lifelong ballers. I couldn't stand that, so any edge I could get, I would take.

In this context, avoiding cigarettes and reefer seemed an easy choice. And so, when I wanted to abstain, I always had the out that I was worried about my wind on the court. To be cool, of course, I wasn't always completely abstinent and I certainly wouldn't preach about not using. But as a result, my early drug use was mainly symbolic and I carefully monitored any high that I experienced in order to avoid feeling like I was out of control.

As is the case for most people, however, the first drug I ever tried was cigarettes, sneaking a stolen Kool or Benson & Hedges with Amp and Mike in my aunt's backyard when I was seven and they were ten and eleven. None of us really knew what to do with a cigarette. Our main goal was to look older and impress the neighboring girls who were hanging clothes out to dry in the backyard that faced ours. Thinking I was safe from adult eyes, I got one from my cousins, lit it up, and inhaled deeply. I blew the smoke out, then posed with the cigarette between my fingers, doing my elementary school best to look Hollywood-cool and sophisticated. Stifling a cough, I found that it only made me dizzy. It also provoked the most excruciating headache I'd ever had, which is actually one of the most consistent toxic effects of nicotine.

Worse, soon, those girls were laughing at us—and not with us. We'd thought that the tool shack, which blocked the sight lines from the house, would give us cover from adult eyes. We even felt like we were making some progress with those ladies, flirting over the fence while we tried to look like men with our "squares," which was what we then called cigarettes. But either my aunt's boyfriend Cooper had noticed that some of his smokes were missing or something else caught her attention. They both came out of the house, very quietly, signaling to the girls not to let on that they were behind us.

Before we even knew what was going on, they were screaming at us, "What in hell do you think you're doing?" and chasing us around the yard. The girls were barely able to contain their hysterics. I never tried another cigarette until I was in the air force in the United Kingdom—and even then, was never more than a social smoker, for the same reasons that drove my moderation with marijuana, primarily concerns about athletic performance. I have never purchased a pack of cigarettes for myself in my life, but during my military service, I did smoke with friends at pubs to enhance the alcohol buzz. I felt that this intensified the excitement that the first drink stimulates. Later, I was intrigued to find a study that examined this phenomenon, suggesting that I was correct.

My first alcoholic drink had been less eventful than my first cigarette. I was probably twelve. I remember going to the refrigerator, desperately thirsty after playing football in the stifling heat. Other than water, the only beverage in the fridge was a pink Champale (the poor person's champagne) and I wanted something better than water. I drank down the whole twelve-ounce bottle, thinking I was enjoying its cloyingly sweet taste.

But what I later realized that I really liked was the sense of relaxation, that calm but also somehow exciting chill that came over me. Again, though, alcohol never became something I needed or even particularly wanted. Street lore had it that twelve or sixteen ounces of the malt liquor Private Stock would keep your manhood erect forever—so I tried that from time to time when I was with a girl. Of course, like most lore, this too was an oversimplification. Sure, a low dose of alcohol can reduce anxieties, and thereby enhance sexual performance. But larger amounts will most likely be disruptive to performance. And so, other than my occasional use of the drug as a sexual aid, alcohol wasn't my thing.

In fact, I was so uninterested in alcohol as a teen that my mother actually kept a full bar including liquor and other supplies in the bedroom I shared with my little brother. She had no fear that we'd indulge. I'd seen how alcohol could make some adults lose their cool and look foolish (I wasn't observant enough to notice pleasant, stress-relieving effects occurring when people drank moderately). I'd also seen how it could make people sloppy and pathetic. One of my mother's friends was a Vietnam vet named Paul. He would frequently show up drunk in our living room and lament his experiences of the war. I felt sorry for him in that state. Mom's alcohol was safe in my room.

Weed was probably the drug I had the closest relationship with during high school. It seemed to be everywhere in the late 1970s and early 1980s (of course, every generation of high school students after the 1960s has said the same thing). But at that point, more than two-thirds of all high school students reported having tried it at least once. In my world, reefer was ubiquitous. Someone in our group always had it. Until I was about fifteen, though, I'd never bothered to smoke it myself. As with cigarettes, I was concerned about potential detrimental effects on my game. But one night, two of my friends— Derrick "Super Slick" Abel and the other I'll call Frank, whom we referred to as Snake—decided that they were going to get me high.

Snake was probably the best basketball player in our neighborhood, about six foot four and two hundred pounds. He was being raised by his grandparents, who spoiled him by giving him pretty much anything they had, as little as that was. They let him drive their old clunker of a car whenever he wanted. Smoking reefer was one of his favorite things to do. And that evening, he and Slick were determined to share the experience with me.

Snake drove us to the spot in Opa-Locka where he bought his stuff. Then we parked at the end of some deserted street and together smoked a couple of joints, listening to the mellow sounds of *The Quiet Storm* on 99.1 WEDR.

"Shit, I don't feel nothin'," I declared. "This ain't shit."

Snake and Derrick looked at me and then at each other. Laughing, someone said, "Yeah, he fucked-up." I continued to insist I was fine and that I really didn't feel any different from usual, but both of them just laughed and repeated, "That nigga fuuuucked-up." Everything I said, every time I laughed or just looked at one of those guys only confirmed for them that I was actually high. I still didn't think so.

In fact, I didn't notice anything unusual at all until they dropped me off back at home. My sister Joyce took one look at me and said, "Damn, you must be fucked-up." I'd heard that same line earlier. I brushed her off. But I think I must've been acting a bit cautious and tentative, not like my usual bold self. My eyes were probably red or maybe I reeked of weed. I didn't yet understand how marijuana affects consciousness.

I went into my room and then things started getting strange. I put on some music and tried to fall asleep. But suddenly, I felt like I was inside the beat. I thought to myself, "What the fuck is this?" The song was surrounding me, throbbing, inescapable.

That wasn't the way music was supposed to sound. My heart seemed to have sped up, too. I felt as though it were keeping time with the R&B rhythm. Was it unhealthy if it did that? Could it kill me?

It was thoroughly disconcerting. I knew I wasn't usually so conscious of my heartbeat; I knew I didn't usually find music so intense. I didn't understand at all that this was what was supposed to be enjoyable. I didn't like having my senses or consciousness altered. I found it disorienting and even slightly frightening—the idea that people would deliberately seek substances that changed the way they saw the world mystified me.

Indeed, I'd never even thought before about the possibility that drugs could change the way you see things. The idea hadn't occurred to me. When I'd watched people get "fucked-up," I'd seen it wholly from the outside, not realizing that from the inside, it could be an entirely different way of experiencing life. All I was aware of was people's outwardly strange behavior.

And as a teenager, I didn't spend much time thinking about how other people saw things; that was part of what allowed me to do things like mess with the white guy on the street. It hadn't occurred to me that perceptions *could* vary much in one person or from one person to another. Later, I'd recognize how understanding the idea of differences in consciousness and changing sensory experiences might let you get a sense of other people's points of view and allow you to empathize with situations that were unlike your own. At the time, however, I was simply distressed by the loss of control. Reefer didn't seem fun or enlightening. If anything, it was kind of disturbing.

Curiously, when I later read sociologist Howard Becker's research on how marijuana users actually have to learn how to enjoy the high, I initially didn't buy it. By that point, I myself had become so caught up in viewing drugs through the prism of how the brain is affected, I'd forgotten the role that social forces can play. Thinking back on my own early experience, however, I realized that I'd been just like Becker's subjects whose first high wasn't memorable or pleasant. It was only when they had smoked multiple times with other users who taught them how to detect and appreciate the sensory distortions and other effects that they began to interpret them positively. Only much later in my career would I begin to recognize that factors like prior experience with drugs and the environment in which they are taken are extremely important for understanding and experiencing drug effects.

During my high school years, however, I just didn't like marijuana. But there was, I soon discovered, a way that I could use the drug to stay on top of things. My cousin Sandra had started dating a guy we called Jamaican Mike, who had a direct connection to a supplier of some of that island's best quality weed. Usually, the Jamaicans and the African Americans didn't mix much in my circle. We looked down on them and vice versa. The same was also true between us and the Cubans and Haitians who were also such a big part of South Florida life. Drugs, however—and sometimes women—could offer some common ground.

Jamaican Mike wanted to be down with me (meaning, considered cool by me) so he always shared his weed. And although I didn't myself particularly enjoy the product, there were people around me whose love for it affected me.

Because I was captain of the basketball team, part of my job was to inspire the other players and ensure that they did their best. Bruce Roy, who was a sophomore at that time, was one of the most talented ballers I'd ever seen. He loved reefer at least as much as he did basketball, maybe more. If we were going to succeed on the court, he was essential. But sometimes he'd miss practice, either because he was off getting high or because of some other drama. I realized that Jamaican Mike's marijuana offered at least a partial solution. Since Bruce was going to smoke anyway, I could supply him myself. That meant he'd have to come to practice if he wanted the best weed.

And that's how I started selling. Again, it was not because of any addiction or even any liking for the drug itself on my part. I did it because of reefer's role in my social world. Weed could get Bruce to practice; I used his desire for the drug to give me more control over my life, ensuring that one of my star players showed up. And although it didn't open my mind in terms of its effects on my own consciousness, it did expand my circle of friends as my access to it put me in touch with more of the school's so-called stoners or burnouts. Before, as a jock, I'd looked down on them. Now I began to see that such people could be cool. Indeed, they turned out to be some of the most open-minded, intelligent, and intriguing kids I hung out with in high school.

I began spending my lunch hour with the school janitor, a brother named Bobby whom I knew from the neighborhood. He reminded me of "Carl the Janitor" from *The Breakfast Club*. We hung with these two white girls, one of whom was a really cool girl named Jana. I'd known her since middle school. She would sometimes get so wasted, taking only God knows what mixture of drugs, that she'd be virtually unconscious in class. Jana had Marcia Brady straight blond hair and wore dark eyeliner.

The four of us would get high at my cousin Betty's house—the same place where I almost got killed sleeping with Naomi, the night that Betty's husband came home and thought I was in bed with his wife. I wasn't sensitive enough to realize that Jana was a lesbian—and that was probably at least part of why I was never able to get with her. But I liked her offbeat personality and never would have become close friends with her if it hadn't been for weed. Experience with the wide variety of people who are attracted to drugs and drug culture would also help me later when I began research aimed at understanding use and addiction.

For those who focus on pathology, of course, my drug experiences would be seen as an aberration. I had many risk factors for addiction in my childhood. These types of risk factors are another part of the dialogue on drugs and addiction that is often misunderstood. For example, I grew up with domestic violence: this alone is linked to an addiction risk that is doubled to quadrupled compared to those who don't live in a violence-plagued home.[5] My father certainly abused alcohol: that's another factor associated with a doubling to quadrupling of risk. Moreover, my mother sometimes smoked cigarettes while she was pregnant with me, and my parents were divorced: both also are strongly linked with elevated risk. In addition, I lived in a poor neighborhood with bad schools at a time rife with racial tension.

With all that against me, you might think that addiction would be inevitable. But that is not how risk factors work. As noted earlier, simply finding a correlation between two phenomena doesn't mean that one causes the other. For instance, a space alien might visit earth and observe a strong correlation between the appearance of umbrellas and the amount of rainfall. The alien might conclude that the presence of more umbrellas causes more rain to come down. This would, of course, be inaccurate. We earthlings know that it simply means that the more it rains, the more likely people are to use umbrellas to protect themselves.

It could genuinely be true that domestic violence does make children more susceptible to later-life addiction—or it could be that each of those things is associated with a third factor, for example, stress, that causes both domestic violence and addiction to increase, while domestic violence itself has no effect on addiction susceptibility. Simply having one risk factor or even many, therefore, may not even directly relate to addiction itself, let alone doom people to definitely develop it. I myself never got close to being addicted to anything.

And even when I later tried drugs like cocaine, I remained unscathed. Further, the reality is that my experience is actually far more typical of drug use than the dramatic addictions we see on TV, in movies, and in books. Most people who use any type of drug don't get addicted; in fact, most people who try particular drugs don't even use them more than a few times.

Consider our last three presidents: Bill Clinton, who claimed he "didn't inhale" the marijuana cigarette(s) he smoked; George W. Bush, who admitted marijuana use and is widely suspected of having taken cocaine; and Barack Obama, who admitted to using both drugs. President Obama even said that inhaling "was the point" of smoking reefer. Whatever your politics, none of them can be seen as not having reached the pinnacle of power and success.

Their drug use was inconsequential—in large part because they all avoided legal consequences from it. If Barack Obama had come up in a time when the drug war was being waged as intensely as it is now, we probably would never have heard of him. A single arrest could have precluded student loans, resulted in jail time, and completely ruined his life, posing a far greater threat to him than the drugs themselves did, including the risk of addiction to marijuana or cocaine. Even among people at the highest risk, like I was, it is still the case that the majority do not become alcoholics or drug addicts.

We got a hit record, they gonna come out for that," Russell Simmons told my brother-in-law Dr. Love, arguing that we should charge five-dollar admission rather than the two dollars we usually got for a Saturday night dance. Russell was managing his brother's group, Run-DMC. He would ultimately, of course, become one of rap's biggest promoters, parlaying Def Jam Records and other hip-hop ventures into a multimillion-dollar fortune. And Run-DMC—with Russell's younger brother Joseph "Run" Simmons, Darryl "DMC" McDaniels, and Jason "Jam Master Jay" Mizell—would soon be the pioneering voices of hip-hop, taking home its first ever gold record and bringing the music into the mainstream. In 1983, though, all they had was a single: "It's Like That," with "Sucker MC's" on the B side.

At the time, rap was still nascent. It was so below our radar that I barely even mentioned to my friends at school that we'd be performing with Run-DMC at our next show. We certainly weren't convinced that people would pay five dollars to see rappers, even if they did have a hit single. We still thought it was kind of uncool, perhaps even clownish. No one had a clue that Run-DMC would amount to anything.

Russell had contacted the Bionic DJs because he wanted his group to tour South Florida—and we were known as the hottest DJs in the South Florida scene. Run-DMC didn't have their own traveling equipment yet, so they wanted to borrow ours for that portion of the tour. We worked out a deal where they could perform with us, using our equipment in a trial run at Washington Park Gym, where I'd attended my first dance in middle school. It wasn't our best venue. We'd had problems with turnout there at times, but it was big and available at the right price and time.

Dr. Love raised our objections to the price but ultimately agreed to Russell's terms. We confirmed the date. And soon we learned from the rappers that the heavy bass beat came from an 808 drum machine. We wanted to see it—but they hadn't even brought it with them. As they performed, we discovered that they'd decided to use the sound from their own record, not the 808, when they played live with us. That left my brothers-in-law decidedly unimpressed. At nine thirty or ten the night of the show, we all went out back to smoke weed before we got started. Someone lit up a fat one and it was passed around as we talked about music and equipment and which of the girls who passed by were the hottest.

As we'd predicted based on the price, however, only about a hundred people actually turned up to see them. The show itself was interesting: I watched as Run's performance captivated Amanda, a girl I'd once dated. I thought, Hmm, maybe there is something to this rap thing, maybe this guy has some talent, maybe this stuff could impress girls. It was hard to believe, but there I was watching her watching them in their black hats and jeans. Run seemed to have her starstruck. Still, the turnout soured the deal for my brothers-in-law, who nixed any future collaborations because they hadn't made much money.

A few years later, in 1986, when I was in the air force in England, I paid to see Run-DMC when they toured Great Britain for the release of their album *Raising Hell*. It started getting major airplay all over the world. And when I returned home the following year while on leave, I found that rap had blown up. Every party, every night, you'd hear LL Cool J's second album and Run-DMC, everywhere.

I saw Run-DMC in interviews telling kids to say no to drugs and staying in school. I had to laugh, remembering those brothers smoking weed with my friends behind Washington Park Gym. But it would still take some time before I could consistently distinguish between truth and bullshit about drugs.

CHAPTER 7

Choices and Chances

Chance favors only the prepared mind.

—LOUIS PASTEUR

caught this fucker stealing," the overweight white guy told his boss as I vehemently denied doing any such thing. I was at a garage/auto supply store. I had already taken four car batteries and loaded them into Derrick Abel's car when I was seen trying to take one last battery toward the door. Realizing that I'd been spotted, I turned around and told the disheveled-looking mechanic that I had a question about this particular battery, hoping he'd think I intended to buy it. The young man responded that he needed to get his manager to answer my question. Then he'd led me to his supervisor, springing the trap on me. He tried to grab me and I knew I had to get out of there, fast. I dropped the battery and bounced.

As I sprinted away, Derrick had already rolled out ahead of me. He knew that the out-of-shape store employee had no chance of catching me so he didn't want to risk slowing down

to pick me up. Seeing no better exit, I scaled the fence that surrounded the parking lot. I was on Hallandale Beach Boulevard
just outside of Carver Ranches, an area that was a mix of small
businesses and homes. The employee—whose belly was hanging
over his belt—took off after me.

But I was an athlete, in peak condition. I raced through the
next yard. I knew that getting caught could ruin my life. It would
almost certainly get me kicked off the basketball team, even if I
wasn't convicted or locked up. The guy continued to follow, trying to catch up. He ran as hard as he could, but he was puffing
with effort from unexpected exertion.

The next yard I jumped into, I realized belatedly, had several
vicious dogs in it. Their loud barking made my heart only pound
faster. I could see their eager eyes and menacing mouths. Trying
to keep myself calm, I looked for the best way out. Racing across
the grass and barely dodging clotheslines and palm trees, I managed to climb the next fence. The dogs were closer to me than the
man was, but neither of them was going to be able to catch up as
I shimmied up the fence.

My hands were getting scraped but I felt nothing. The hounds
continued loudly snarling as I made my way toward Twenty-Fifth
Street. The guy from the shop was now nowhere to be seen. I had
lost him at the first fence. But I was sure that a call to the police
had been made by then. I wasn't sure if they were for me, but I
could hear sirens in the distance. They seemed to be getting louder
so I kept running. Inside, I was laughing at the fat man, but I knew
that if I did get arrested, the consequences could be serious.

Soon, however, I caught a break. My friend Reggie Moore,
whom we called Tudy, happened to be driving by and saw me
racing down the sidewalk. I flagged him down. I was dripping
with sweat. He stopped just long enough to let me in to his '72
white Buick Skylark and rapidly drove off. As we got farther

away, I began to relax and my heartbeat slowed to a more normal pace. I laughed as I realized how lucky I'd been and eased down in the front seat. I shuddered at the chain of coincidences that had made my escape possible. I don't know if I'd ever been happier in my life to see someone.

During my last two years of high school, I'd become increasingly involved in ever more serious crime. None of it was violent; all of it was calculated to minimize risk while getting extra money beyond what we could earn at our minimum-wage jobs. My friends and I regularly stole batteries and rims from cars and sold them to garages and gas stations. And earlier, during my freshman year, I'd started hanging with some kids who burglarized houses.

By then, my family had moved to the projects, which were in Dania. Since most of my friends were still back in Carver Ranches, however, that's where I spent most of my time. Sometimes I'd stay with my girlfriend Marcia, Big Mama, or Grandmama; alternatively, I'd try to get a ride home or just hang out all night on the corner.

My cousin Larry; a guy known as Pink, who was light-skinned enough to be mistaken for white; and one called Dirty Red, who had freckles and red hair but was a bit darker: these were the brothers I hung out with at that time. We'd hang at the intersection of Twenty-Sixth Street and Forty-Sixth Avenue; the neighbor folks called it Junkie Corner. But it wasn't what you might think: no one shot heroin, nodded out on their feet, or sold smack there. It was just the spot where young men drank Private Stock and smoked reefer. It was also where we bragged about our sexual conquests and made half-assed plans to steal TVs or other property from unsuspecting white folks.

"Yo, I know some people who are out of town; let's go to their crib and get some shit," someone would say.

"Yo, you down?"

"You know I'm down."

"I'm down," everyone else would say.

"Cool," we'd agree, and then pile into two cars and roll to the white section of town as if no one would notice us. I'd always stay in the car. If we'd been busted, I now realize, I'd have been considered the lookout, but I didn't think of it that way. Sometimes, I was just trying to get a ride home. Other times, I'd get a share of the loot, like a camera that was smaller than my hand, which was probably extremely expensive back then.

I always tried to be alert to the potential risks as well as the benefits of the crimes I committed. Though it may have looked like teen impulsiveness (and, of course, I did have the adolescent cockiness that creates risk-blindness—pointing the gun at that white man wasn't exactly a smart move), I wasn't usually stupid, either. I wouldn't do things that I'd seen people catch a case for; I wouldn't risk shoplifting at that mall filled with cameras and security guards and I wouldn't do anything violent like mugging people. My goal was to stay in school so I could become a professional athlete.

Once, while the guys were burglarizing someone's home, they had to fight off some girls who came back unexpectedly and caught them. But fortunately, that was the closest I ever got to getting into trouble with those guys. We laughed it off, not even thinking how our behavior might have affected those girls. In fact, we mercilessly teased Larry, who had punched one of them while trying to get her purse. He'd hit her so lightly she didn't even drop the bag—and then he'd had to run to the car before we drove off without him.

As with my earlier lawbreaking, these activities had nothing to do with drugs and everything to do with street credibility. Even as I participated in burglaries and stole batteries, I also worked whatever job I had. I diligently showed up when required and always did what needed to be done, not seeing any contradictions in my behavior. I worked hard because you were supposed to work hard; I stole because there was never enough money; I went to school so I could get a basketball scholarship. At sixteen, I still thought I was going to play in the NBA, though earlier, the dream had been the NFL. The main career plans I ever had as a kid were these hazy visions of becoming

Standing in the hallway at Miramar High School during my senior year.

a professional athlete. Fortunately, they had the side effect of keeping me in school.

I also felt justified taking from those we viewed as having excess, like we were Robin Hood. My highest paid job in high school barely earned me four dollars an hour. (Though the older guys made money from deejaying, I was just glad to be up front and part of that scene with my brothers-in-law. I got my money elsewhere.) When I later learned about psychologist Lawrence Kohlberg's stages of moral development, I felt vindicated. I'd reached the "highest" level of moral thinking, according to him, in early childhood: I'd gone from thinking that rules alone determined what was moral to thinking about universal principles of justice, before I'd hit my teens.

It had always seemed obvious to me that if, say, your family needed a lifesaving drug, it would not be immoral to steal it. What kind of person would let arbitrary rules that let rich people have access and let poor people die stop him if he had a choice? I didn't understand why everyone wouldn't see the situation as unjust if property was more valued than life.

During my senior year, Derrick Abel and I plotted with a guy we knew who transported money from a local movie theater to the bank. We were going to rob him but not hurt him; he was, in fact, our inside man. The runner, we were told, carried thousands of dollars. It would be our biggest heist ever. We talked and talked about it. Our friend Alex, however, refused to get involved. He was about five foot eleven, with a small mustache and muscular build. I'd always thought he was cool. But he said, "Fuck that shit. That's stupid." To my shock, he flat out said no.

Thinking back later, I realized he was from a two-parent family and had had a lot more guidance than I did. At the time, though, we decided at that instant that he was uncool. Fuck him, we're no longer friends. We dropped him without further thought for a few weeks; someone that punked out couldn't be down with us; he couldn't be trusted. I didn't see this as cold or callous; that was just how it was.

In fact, it boggled my mind that someone would ever say no to his boys; for me, cool and its requirement of loyalty to our group always came first. It was the foundation of my values, one of the few things that really meant something to me and structured my social life. Putting those ties at risk, to me, seemed much more dangerous and threatening than anything the system could do to you if you ever did get caught. If you stayed cool, you could handle that. If not, you weren't a man and there was nothing much to live for anyway. As it happened, we never got around to robbing the guy. I reinstated my friendship with Alex about a month later. But I never shared information about my capers with him again because I knew he wouldn't be interested in participating.

Episodes like my narrow escape from the battery store and our somewhat arbitrary decision not to do the movie payroll job pose deep questions about the role of luck and chance in a person's life. If we'd gone ahead with that risky plan or if I had gotten caught and punished for some of my other activities the way so many of my friends eventually did, many of the opportunities that I have had would almost certainly have been lost to me. It wasn't that I didn't do the foolish things that other kids around me did; it was that I didn't get caught doing them. Like Presidents Obama, Clinton, and George W. Bush, some of my fate rested on not getting caught taking drugs or engaging in other "young and irresponsible" activities.

As a scientist, I'm familiar with Louis Pasteur's notion that "chance favors only the prepared mind"—the idea that while luck does play some part in great discoveries, hard work prepares the soil without which they cannot grow. The same is true in my life. Without a lot of hard work, I'd never have gotten to be where I am. Unlike luck, hard work is under your control: you can either do it or you can take shortcuts. That's quite clear and often differentiates between winners and losers. I believe deeply in putting in the effort and tell my children so ad nauseam.

But I'm also acutely aware that often, hard work isn't enough, especially when the stupid things that black children do are punished much more severely and with much more lasting negative effects than happens with the equally stupid things that white children do. Of course, I'm not arguing that crimes like robbery and burglary shouldn't have consequences. They should. I just think that the consequences should both be educational and allow for redemption.

And data shows us that the criminal justice system is not the best way to impose these consequences. Its personnel aren't trained as educators or counselors; they're trained to contain damage and dole out punishment. Besides this, prisons are difficult to run in a way that keeps children safe and healthy and they are far more expensive to operate than alternatives that are actually more effective. It's not just my experience—or that of our last three presidents—that suggests that avoiding the justice system produces better outcomes. This is clear from multiple studies.

This data shows that teens who are either not caught or are given noncustodial sentences for their crimes do much better in terms of employment, education, and reduced recidivism than those who are incarcerated or otherwise removed from the community and grouped with criminals.

One large American study examined the cases of nearly one hundred thousand teens who had their first contact with the juvenile justice system between 1990 and 2005. Fifty-seven percent of these youths were black; the overwhelming majority were male and their average age was fifteen. Most had been arrested either for drug crimes or for assault; all were studied at the time of their first offense.

The researchers found that, regardless of the severity of the initial offense, teens who were incarcerated were three times more likely to be reincarcerated as adults[1] compared with those not incarcerated for similar offenses. Being locked up hadn't deterred them; rather, it had forced them to spend time with criminals, had possibly taught them more about how to commit different types of crime, and ultimately set them up to be reincarcerated.

Similarly, Canadian researchers conducted a large-scale, carefully controlled study in which 779 low-income youth in Montreal were followed from ages ten to seventeen; they were interviewed as well as their parents and teachers. Years later, researchers examined their criminal records and found that those who had received any kind of custodial sentence as teens were thirty-seven times more likely to be arrested in adulthood than their peers who had committed similar crimes but were not incarcerated during adolescence.[2]

The data from the above studies and others clearly shows that segregating troubled teens together in settings where there are no parents and few peers aiming for athletic or academic success tends to make their criminal behavior worse.[3] Both being labeled as a "bad kid" and hanging out with peers who feel that their only source of manhood and identity is engaging in criminal behavior significantly increase risk for future crime. Social influences like incarceration during youth predict adult crime far more strongly than anything we've been able to identify so far related to biological factors like dopamine in the brain.

Moreover, because black youth are more than twice as likely to be arrested as whites,[4] the negative effects of juvenile prison have a disproportionate effect on our community. (For drug offenses, the inequities are even more glaring: drug cases are filed against black youth at a rate almost five times greater than for white youth, even though more white youth, 17 percent, report having sold drugs than blacks do, 13 percent.)[5] While these facts are discouraging because they show how big the problem is, they also suggest that a clear solution is minimizing juvenile incarceration rates.

The lives of my friends, neighbors, and relatives showed this contrast clearly. Those who managed to avoid contact with the system, as I was, were much more likely to make it out of the hood and into the mainstream. Meanwhile, many of those who got caught never recovered, even if the first offense was truly minor. That one incident would lead to increased scrutiny and then further arrests—or the exposure to juvenile detention or other incarceration would harden a criminal identity and/or connect people to those involved in more serious crime. It was as though a pebble had set off a rock slide. A small event produced a chain of devastating consequences, forever altering a life course.

One of the saddest examples of this in my life is the story of my cousin Louie. The math whiz pitcher with whom I'd shared a bed at Big Mama's had been an honors student when his mother switched him from one high school to another. Once he got there, the small, skinny kid felt that he had to prove he was down with a new set of friends.

Shortly after his school transfer, Louie was picked up by the police for truancy or some other trivial, nonviolent offense. For that he was sent to juvenile detention at fifteen; the few months he spent inside hardened him and gave him the reputation he'd been seeking, rather than serving as any kind of deterrent. Having survived being locked up, he saw himself as one bad dude. Rather than returning to his advanced math classes, he skipped

more and more school and started hanging out with the professional thugs. Soon he dropped out entirely.

By then he was pulling armed robberies, jacking trucks carrying radios, TVs, and other electronics and appliances. He
and his friends once hit a Brinks truck and successfully hid
the money so well that it has yet to be found. But the rumors
about that heist marked the peak of his glory. In his mid- to
late teens, he began drinking heavily and smoking weed, and
by his early twenties, he'd started smoking crack. He ultimately
spent at least ten years in prison—and now lives in a halfway
house, barely able to function on the psychiatric medication he
was prescribed when he entered prison. Although the details are
unclear, it is said that the medications were originally prescribed
to control his anger.

Fortunately, there are also positive life events that can lead
to spiraling virtuous circles, not escalating vicious ones. For
me, one of these was my decision to take the Armed Services
Vocational Aptitude Battery (ASVAB). Although I'd worked
relentlessly at athletics and had big dreams about college basketball and the NBA, I'd otherwise given little thought to what
I'd do after high school. Since I'd told all my friends that I'd
be getting a big college scholarship, I knew I had to leave home
somehow—or risk losing the rep I'd so assiduously worked to
build up.

I knew nothing about how college basketball really worked
and the importance of coaches in getting scholarships for their
players. I was ignorant of the machinations and realities of that
world. All I did recognize was that without a full scholarship, I
probably couldn't afford to go to college. I needed other options.
I wasn't likely to get much financial support from my mother. In
fact, I figured she'd probably pressure me to stay home and work
rather than encouraging further education. In our family—as in

many others in my neighborhood—children were expected to support, or at least partially support, their parents once they reached working age.

My father wasn't going to be of much use, either. He'd never demonstrated that he had that kind of money to spend on his children. Although I sometimes saw him around, by this time we had drifted apart as fathers and sons often do during adolescence. The thought of having to depend on my mom for college funds or the notion of skipping college—and the chance of a basketball career it offered—and going to work full-time was not appealing to me.

Maybe these considerations were in the back of my mind; maybe they had nothing to do with my decision to take the armed forces test. All I remember is that early in my senior year of high school, I decided to take the ASVAB because it meant that I didn't have to go to my classes that day. I know for sure that I had no active desire to join the military. The guys I'd seen coming back after they'd joined the army or the Marines seemed brainwashed, no longer concerned with what we cared about and valued. But my guidance counselor, Ms. Robinson, had said I could leave school early if I took the test—and I knew I could bubble in the answers quickly and be on my way to hang out with my friends much faster than I could be if I went to class. That nearly random choice had a considerable influence on my life.

In the school cafeteria, faced with a number-two pencil and a question booklet, my primary goal was to get done quickly. I didn't fill in the little ovals at random, however. That seemed dumb, even though I told myself that I didn't care about my score. I did guess without much thought or leave questions blank if something didn't come to me easily, particularly on the reading and vocabulary sections.

When I got to the math section, however, I found myself pay-ing real attention. I had my pride. I thought to myself, you might trip me up with English or social studies but not math. I did my best on the math sections of the exam. Then I turned it in and forgot about it in my daily routine of basketball, spending nights with my girlfriends, and rocking the mic on the week-ends. I didn't give it another thought.

A few months later, the results came back. To my complete astonishment, I was told that I was one of the only people in my high school to score high enough to be recruited by the air force. At the time, it filled me with pride. Now, however, I don't think that this shows that I was especially smart: the college-bound kids didn't take the ASVAB and I suspect I wasn't the only one who'd simply taken it to get out of class. The scores would have

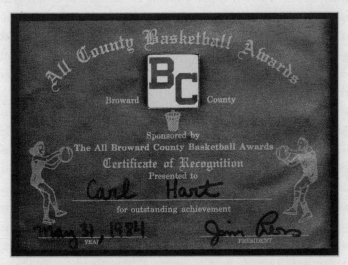

Despite receiving all-county basketball recognition, I didn't receive a basketball scholarship. As a result, the military became a more likely option.

been much higher if the whole class had been required to take it—or if only the college-bound students did so. It didn't represent a true picture of the smartest kids in the school.

Today, I'd call that a sampling bias. For example, in my research, I have to think about not just the drug I might be studying but also the kinds of people who would be available for study participation and whether they fairly represent the people I'm trying to understand. While I explore their subjective experiences with them, I also study their behavior on different days and with different doses of drugs. These contextual factors matter a lot: under one condition, I might find one effect but under another, I might find the opposite outcome or no effect at all.

I often explain it this way: Imagine if the only experience you had of driving was being plunked down behind the wheel for the first time in your life in a raging thunderstorm or blizzard on a busy highway. You'd probably think driving was seriously dangerous and that most people couldn't handle it. You might generalize from your own single experience, under those awful conditions, to that of everyone and see driving a car as something that should be highly restricted.

Of course, your sample of driving in that kind of situation is limited to one sample of an extreme situation. It doesn't include driving on a bright sunny day, driving once you've had years of experience, or driving on a quiet country road. Similarly, using a drug once or twice or seeing a friend become really paranoid as a result of that drug does not provide an adequate sample of the range of possible drug experiences. Likewise, sampling only the non-college-bound students' results on a test of intelligence does not provide a representative sample of possible test scores for a particular high school class.

But learning to think about ways to really isolate the causes and effects of things was one result of that rather random choice I made to take that test. It eventually opened up a whole new world to me. If I hadn't made that one, seemingly irrelevant decision to take the ASVAB, it's unlikely that I would now be a scientist and college professor.

Once those results were in, however, both the army and the air force did a full-court press to try to recruit me. At first, I didn't really take it seriously. My guidance counselor nonetheless insisted I meet with both recruiters. She set up the meetings in her office and got me out of class for them, once again successfully ensuring my attendance by understanding what motivated me. Although I either acted like a clown or literally slept through many of my classes, Ms. Robinson found me charming and didn't give up on me, knowing that the military was one of the few options that could make a real difference in my life. Her exceptional dedication to trying to secure a future for me really mattered.

I continued to be quite resistant at first. One of the most depressing experiences I'd had as a child was listening to our family friend Paul talk about Vietnam. He was invariably drunk, reeking of alcohol. His memories seemed overwhelming: he'd suddenly start regaling us with stories of seeing men's heads exploding, their faces blown to bits. His expression of horror and the physical manifestations of fear like flop sweat illustrated much more than his words how the experience of war had broken and devastated him. He'd talk about friends who were crippled or dead, and other brothers who came home physically whole but were no longer all there mentally. He warned us over and over not to sign up, that black men were even less valued by America when we were sent to war. I wanted no part of it.

Of course, as recruiters do, the army and air force representatives painted a very different picture. Emphasizing basketball and college study, stressing that the country was at peace, they glossed over the main job of the military. No mention was made at all of war or combat. I didn't have to worry about that. They implied that all I'd have to do was take a few orders and stay physically fit. They explained how in the military—unlike in college—not only could I play ball; I could get nearly free college tuition, too. They praised my intelligence and my skills and kept the focus on what was in it for me.

As I saw it, my only other alternative was to apply for financial aid, but I had no idea how I'd be able to raise the rest of the money to pay for the full tuition and room and board. The idea of continuing to be dependent on my mother distressed me. And I knew I couldn't stay home and face the disappointment of my sisters and Big Mama, who had cheered on my athletic career and encouraged me to stick with school. I certainly couldn't handle the smirks of my rivals if I didn't leave Miami to play college basketball somewhere. And so, before long, I found myself no longer considering whether to sign up but trying to decide whether the air force or the army was the better option.

Again, a random choice—one that might seem completely unlikely—put me on the path to my future. I met several times with each recruiter. The army recruiter was a brother. He tried to sell me on his branch of the armed services by demonstrating how cool he was and by extension, how cool I could be if I joined the army. As you know by now, this normally would have sealed the deal with me, but I didn't buy it from him. I felt that he was trying too hard. His behavior wasn't authentic; he seemed like a fraud to me.

In contrast, the air force recruiter was a classic white dork. He made no attempt to be cool or to pretend that he was anything like me. Instead he was straightforward and made a plainspoken pitch. He understood intuitively that he'd never be able to impress me by trying to be something he so obviously was not—and that in itself made an impression. It made him seem trustworthy.

Still, I kept considering both branches of the military. I may have inadvertently fallen for one of the oldest behavioral tricks there is. That is, being presented with two options when I'd initially wanted none of them, but then seeing my choice as one of those two and forgetting any other possibilities. At one moment, I found myself staring at the army green uniform and thinking, I can't do that, can't do that. That shit's not for me. It offended my sense of style, somehow. I couldn't picture myself dressed that way, ever. Then I'd think about basketball and scholarships and think, Maybe.

Later, at one meeting with the army recruiter and his superior officer, I actually fell asleep because I'd been out so late the night before with a girl. Falling asleep in class wasn't uncommon for me, but that was the first time I'd done it in a small meeting. The guy started pressuring me because he said my napping had embarrassed him in front of his boss, so I should sign up to make it right. But then I started thinking again about the horrible green uniform and what the air force might be like instead.

Finally, I went back to the air force recruiter. I had come to associate the army with some of the less intelligent brothers I knew: that was the branch of the service they seemed to join. The air force had an advantage here, especially since I remained flattered that I'd attained the higher intelligence score needed to join. Their uniform wasn't entirely intolerable, certainly not as awful as that army green.

The airmen were sharper, in both mind and dress. It sounds a bit weird thinking back on it, but again, another not-totally-considered choice—preferring the blue of the air force to the army green; wanting to be part of a service that required a higher IQ—put me on the path to science.

Because I was still only seventeen, my mother had to sign off on my enlistment as well. It was an ironic moment for me. MH was sitting at a table at Grandmama's house, with all the paperwork in front of her. The air force recruiter was there, showing us how to fill out the paperwork. Suddenly, she paused. She looked up at me before she finished signing and asked, "Are you sure you want to do this?" Remembering all the times she hadn't been there when I needed guidance, I thought to myself, Oh, now you wanna play mommy? Just sign the fucking papers. I felt as though she was only feigning maternal behavior to impress the recruiter.

CHAPTER 8

Basic Training

Don't try to change yourself; change your environment.

—B. F. SKINNER

The military has its indoctrination down to a science. They know how to use experiences like exhaustion, peer pressure, isolation from one's friends and family, and disorientation to maximum effect in boot camp, or basic training, as it's formally known. Even though the physical challenges were nothing compared to the workouts I'd already done throughout high school, the mental challenges to my ideas about myself, about race, about self-control, and about what I wanted were immediate and at times, daunting. I started on August 24, 1984.

The night before I left, the air force had agreed to pay for a hotel room near the airport so I would be sure to be on time for my early morning flight to Dallas. I stayed up nearly all night with my high school friends, knowing it could be the last time we'd really get to hang. We were laughing and joking, the guys telling me that I'd come back all "shot out," or brainwashed like other guys from the neighborhood who'd joined the service. But I wasn't anxious at all until the morning came and I headed for the airport. It would be the first time I'd ever flown.

Although it would have been just as easy for the military to fly us directly to San Antonio, we were sent instead to Dallas, where we had to wait for hours at the airport. Then we took an extended bus ride to Lackland Air Force Base.

It's ingenious because the exhaustion starts to wear you down before you even know it. When we finally got to Lackland, it was about midnight. And it still wasn't time to rest. For what seemed like hours, we were made to stand at attention, the boredom and physical stress of the position draining our minds and our bodies. There were no clocks and, of course, not knowing what time it was added to the discomfort and disorientation.

At some point, the training instructors came out yelling. Hurling abuse at us and calling us pathetic mama's boys, they began the next phase of our indoctrination. I thought to myself, This is a fucking joke, and almost laughed because it was so much like every clichéd boot camp scene I'd watched in movies like Private Benjamin and An Officer and a Gentleman. There they were, like drill sergeants out of Central Casting, ridiculing our dress, five o'clock shadow, and overall lack of competence.

Soon they targeted one of the biggest men among the recruits for an extra heaping of humiliation. He was a white guy, huge and incredibly built.

"You want to do something?" one of the instructors said.

"No, sir," he responded.

"Why the fuck you looking at me? You calling me a liar?" And on it went.

I knew right then that I'd never be the same. There were three instructors, all at least as powerfully muscled as the most fit recruits and full of pride. They came at him like they were going to kick his ass, getting in his face as he stood there sweating. The guy knew he couldn't fight back, so he tried to respond as submissively as possible. Then one of the trainers said to one of the others, "Sergeant Castillo, hold my shit. I'm gonna fuck this motherfucker up!" The man stiffened, unsure what to do. By the end, he looked like he was close to tears.

Watching them push him to see if he would snap, I knew that I had a choice of my own to make about how I would behave. I could buckle down and do what I had to do and maybe get something out of this, or I could be a clown and continue aimlessly, taking nothing but sports and street status seriously. I could let these authorities defeat me by dropping out or I could be serious and stay. I thought about my sisters back home and I didn't want to let them down. They had seen the military as a new start for me and as a way out of the dead-end jobs that seemed destined to be my future otherwise. Along with Big Mama, they had encouraged me and mothered me, placing so much of their hope for the future in me. I couldn't stand the idea of disappointing them.

Although I still had big dreams about basketball, somewhere inside, I knew that fully grown at five foot nine, despite my talent, the odds were against me having any kind of professional career. If I was going to make something of my life, it had to start here and now and I had to have a different attitude. I wasn't going to let any of those sorry, out-of-shape recruits I saw in my squadron do better than me. Luck may have helped me get there, but that epiphany and my own hard work that followed were what allowed me to take advantage of the opportunity. There would be many chances for me to fall down and be pulled back— but the first day of what we were told to call "basics" turned my head. We were all relieved when the instructors dismissed us and we finally were allowed to get a couple of hours of sleep.

And once I resolved to put in the effort, there was not much else to do other than submit to the experience and do the work. Although most people find the constant exercise in boot camp to be physically grueling, I found that I faced a different challenge. At home, I'd played a minimum of several hours a day of basketball in games and practice, constantly running and doing specific drills to keep my edge. That didn't include pickup games and other athletic activities I did just for fun. In basics, we were being trained so that after six weeks, we'd be able to run a mile and a half as a squadron. And we had to go at the pace of the slowest man, who was seriously slow.

To be fair, it was San Antonio, Texas, at peak summer heat and not everyone had grown up in Miami and become acclimatized to intense exercise in high temperatures. But I felt like I was teasing my body. When we were done working out, I'd barely even be warmed up. As a result, I wound up running push-up and sit-up contests with my bunk mates at night: I told the guys that we could all get out of here looking good if we made some additions to the routine.

Back home, brothers who had spent time in prison would typically return looking amazingly buff. They said that in the joint, they'd done those exercises constantly—so I reasoned that we should do the same in the air force. Pretty soon almost everyone in my squadron was doing it. We'd take bets on who would win with the highest numbers.

The only other thing to do at night was write letters home, which became another way to compete. The more letters you wrote, the more you would get back when the training instructor handed them out at mail call. Receiving lots of mail was a sign of high status. I wrote to all my girlfriends, as well as my sisters and brothers.

And as with its use of psychology to break us in with exhaustion and boredom, I found that the air force was far more adept than I expected it to be at dealing with racial issues. In their history of how the army (and by extension, the rest of the military) became the most integrated institution in America, *All That We Can Be*, sociologists Charles Moskos and John Sibley Butler wrote that the service is "not race-blind, it is race-savvy." That's how I felt about it. The air force had been the second of the services to desegregate and was the first to become fully integrated.

I was amazed by how quickly the military got everyone—black, white, yellow, brown—to work as a coordinated unit. They imposed rules to ensure we would get along and by giving us the common enemy of the training instructors and their strict command, united us in a shared experience. That created a bond. I first got a real sense that things worked at least somewhat differently in the military when I saw our dorm chief, a black guy, get demoted for giving favors to some of the guys in our squadron. Someone had dropped a dime on him (snitched)—a black guy. It simply blew my mind that a brother would give up another brother: when you grew up where I did, that just was not done in any setting that had real-world consequences.

Of course, the idea of being loyal to a mixed-race team wasn't new to me—that had been part of athletics for virtually my whole life. Off the field, however, I'd always found that those allegiances were not as strong. Race was still foremost in people's minds when it really came down to it. No one I knew believed that American institutions could really be fair to us. We'd all seen people who had faith in that get violently upended, whether through experience of police brutality or employment discrimination or just daily experiences of lack of respect.

There were peculiarities and misunderstandings, too: for example, the term *homeboy* was banned after white guys mistook it for an insult. They thought we were using it to demean people, to say they were homebodies who never went out and were antisocial. Of course, we were actually referring to friends,

Basic training photo.

particularly people from our neighborhood that we liked. But it made whites uncomfortable, so it had to go.

Still, such incidents were not as common as they were in the civilian world and overall, I felt like we were treated with respect based on how we behaved, rather than on race. The military rules were clear and felt less capricious. I began to change my attitude and become more open and hopeful about the future.

I have to emphasize here, however, that I did not change overnight.

There was nothing sudden at all about my transformation

from a kid with a poor education who knew little about the history of my people and the ways of mainstream America into someone capable of becoming a tenured professor at an elite university. I only gradually became aware of the gaps in my knowledge, and the analysis that I undertook would allow me to transcend them by understanding their roots and the forces that shaped my family and neighborhood. I didn't go instantly from being an indifferent student to one who spent hours in the lab. And I certainly didn't change from someone whose focus was primarily on my social life into a serious academic simply by joining the air force.

But the air force was the environment that allowed me to start to make these changes, to start to understand what I'd missed in my earlier education and my own capacity for change. My commitment to the service made in boot camp was only the beginning. There would be many more times when I failed to make the best choices, when my lifestyle threatened to swamp my desire for a different life and when the pull of the reinforcers I knew was stronger than my commitment to my future. In fact, my college education itself started in part because of a choice I made to take drugs to be cool with my friends.

An imposing Bob Marley poster hung on Mark Mosely's door, larger than life, showing the reggae star onstage at the height of his power. Bob's dreads were arrayed around him as he sang into the mic. The scent of incense—typically sweet jasmine—wafted into the hall from Mark's dorm, but the poster was hung on the back of the door, visible only from inside. His window shades were usually pulled down and the lighting was dim.

When Bob's music was not emanating from a record on the Denon turntable or reel on the Akai 747, Mark played other reggae and jazz. His place looked like a revolutionary Afro-centric pad from the 1970s, but it was actually located in a recently built residence hall in Okinawa, Japan, on Kadena Air Base in 1985.

Several years older than me, Mark was a jet mechanic; I met him because we lived in the same building on the base in Japan to which I'd been assigned after I'd completed my initial training. His goal was ultimately to study sociology back home at the University of California. In the meanwhile, he would school other black airmen to raise our consciousness during our military service.

But—although some associate Marley and his music primarily with the marijuana that is a sacrament for his Rastafarian religion—Mark did not take illegal drugs. He didn't burn incense to hide the odor of cannabis and he didn't lower the lights to shelter eyes reddened by marijuana. The higher consciousness he sought involved intellectual and revolutionary enlightenment.

In fact, the marijuana smoking that led indirectly to my college career occurred in a different social setting. I got high with another group of friends in Japan. It was during this time in Okinawa that I began to realize that I would have to make some real decisions about the peers with whom I surrounded myself because the ideas and habits that I shared with them would be an important influence on my future. Like I've said, I didn't become studious and intellectually driven overnight. Mark was a powerful force in my education, but there were competing elements. At first, it wasn't at all clear that I'd manage to keep my commitment to myself, my sisters, and the service.

I'd originally been slated to go to Nellis Air Force Base in Las Vegas, rather than overseas. But my cousin Cynthia, whose husband was in the air force and stationed at Kadena, convinced me to swap with another new recruit and join her and her family in Japan. I knew nothing at all about Japan or its culture. I did know that a girl I was seeing was going to be sent there. I thought it would be interesting to see another country, and having a friend-girl and at least some family there would make the transition easier. It seemed as good a place to start as any. As far as I was concerned Okinawa was the same as Tokyo, and Tokyo was like any big city in the United States. Boy, was I wrong.

I didn't realize that my cousin had told me only the good things, hoping she'd be able to get me to join her church there so she could save my soul. I quickly made it clear that that was not on my agenda. And I rapidly learned as well that Okinawa had a reputation as being a hardship post for single men, known disparagingly as a prison island and dubbed "the Rock."

It was especially difficult for black folks. Racism in Japan felt more conspicuous than it had growing up in the American South, perhaps in part because I wasn't expecting it. But the Japanese had seen all the American movies and they knew who the niggers were. On more than one occasion, shopkeepers off base actually used that word to refer to me to my face. Even when it wasn't that overt, it was clear that I was being treated like a second-class person in many of my interactions with the locals.

Still, the most disturbing thing to me about Japan was the lack of American servicewomen. So few American women were stationed there that I regretted my choice almost immediately. It was almost as bad as in boot camp, where the men and women were completely separated. Of course, outside the base on Gate 2 Street, everything from tennis shoes to sex was being sold, a plethora of cheap, ephemeral products and pleasures. I had too much pride for that. I wasn't the kind of brother who had to pay for sex.

Even stranger was living away from my family, and all its noise and bustle, for the first extended period in my life. Until we'd moved to the projects when I was in high school, my mother had never had more than a two-bedroom home, which meant that up to six of us siblings—girls and boys—shared one bedroom. My grandmothers' and aunts' homes hadn't been any less crowded and the barracks during basics weren't much different.

Now, though, sharing a room with just one guy was eerily calm to me, especially since the roommate in question probably had what we'd now call Asperger's syndrome. A white guy, he specialized in languages and knew five of them. But although he drank heavily like many airmen, he never wanted to go out. He didn't want to hang out with anyone; he just drank in our room alone. For some reason known only to him, he'd sit there and watch the movie *Trading Places* over and over and over again.

I thought my unease and difficulty sleeping was related to his peculiar behavior, so I got another roommate, a brother I dug a lot. But no, the deep silence of living in a place that wasn't inhabited by a large family, that didn't involve frequent social interruptions, remained mind-numbing to me. It drove me crazy.

Kadena was like a small city, home to nearly twenty thousand American service members, with four thousand Japanese staff as well. Nine hundred miles from Tokyo, it was often as hot and humid as Miami and similarly subject to tropical storms. I'd trained briefly in Denver after completing basics and there I met a guy named Bobby. Bobby was also sent to Okinawa for his first duty station. When we met up in Japan, he, his roommate Keith, and another young airman named Billy were the people with whom I hung out most often initially.

Almost immediately, Keith informed me that he had the

hookup and we all began smoking weed together. It didn't even occur to me not to smoke with them. Being cool still came first for me. I did worry, however, about getting caught by the random urine testing we had to undergo. You might think that this would have deterred me, especially since I wasn't particularly into the high.

But I did care about my social status a great deal. Though from the outside, it might have looked as though I was heedless of consequences, I wasn't. Rather than declining to smoke, I took what seemed in my mind like a logical step to reduce the harm that might come from getting caught: I enrolled in my first college classes.

Ironically, it was my weed smoking that prompted this, not my consciousness-raising friend Mark. My thought was that if I got caught and discharged, at least I'd have a good start on my education before that happened. And that way, I wouldn't let Brenda and my other sisters down so much. Though that was obviously not the intended outcome of the military's drug testing policy, it did turn out to have positive results for me, if only in this indirect way.

So, although Mark influenced my thinking more, it was, oddly enough, the weed-smoking brothers who got me started in genuine higher education. On base, courses were offered by Central Texas College. One of the first classes I took was algebra. I figured that I could build on my math skills and aim for a degree in accountancy or something similar.

This was, I later learned, another example of motivation relying on behaviors that get reinforced and rewarded. I'd been praised and had had early success at math, so I knew I could do it. I'd experienced the pleasure that could come from this myself. My very presence in the air force was one of the rewards for my math aptitude, though, of course, it didn't always seem like one. I also probably chose to take algebra first because I did not want to get discouraged if I tried something new, worked hard, and did not manage to do well. And, as it turned out, I easily got a B.

That gave me confidence when I began to take other classes that I was less sure I'd be able to ace, like Human Resources. For that class, I had to write papers. Though I suspect now that they were pretty bad, I had a friend type them for me and was readily able to get another B. Even in my first year of college, I had no idea that I'd eventually end up as a scientist, studying the complex and challenging human brain itself, no less.

Outside of the classroom, however, I continued to dislike Okinawa. About once a month, Keith, Bobby, Billy, and I would drive up to the top of a hill near the base's high school, with a breathtaking view out over the island of Okinawa. We'd sit in my Honda Accord or in Bobby or Billy's Toyota, smoke, and talk about our plans for when we got back to "the world." We felt as isolated from events back home as we would have been on another planet. Alternatively, we'd go to Gate 2 Street, which was as bustling and chaotic as New York's Canal Street, and around once a week, we'd steal the latest VHS movies to watch on Billy's VCR. As a result, I'm deeply familiar with most Hollywood films of 1984–1985.

The rest of my free time was spent working out or hanging with Mark. He introduced me to a book called *Bloods*, by Wallace Terry, which detailed the maltreatment of black soldiers in Vietnam, in frequently horrifying first-person accounts. That made me think back to the stories I'd heard from Paul, whose memories had seemed so vivid and inescapable. Fortunately, during my time in the service, we were not at war.

Indeed, war was so far from my mind during my time in the air force that the only time I was required to patrol with an

M16 to defend my base, I was outraged. That was later, when I was stationed in England. We had bombed Libya in 1986 in response to a terror attack on a German disco frequented by American soldiers. The planes that refueled the bombers came from the base where I was serving; I threatened to call my congressman about this onerous antiterrorism duty when I was made to do it. Of course, my fellow airmen just laughed. I was lucky not to have faced anything like the combat duty those brothers did.

Mark turned me on to jazz as well. When he played Ella Fitzgerald, I was surprised at his choice of records. I'd always thought that her voice had belonged to a white woman. Mark explained that Fitzgerald's singing might have been dubbed into films starring white actresses, creating that impression and hiding the true source of her glorious sound.

And when Bob Marley sang about freeing our minds from mental slavery in "Redemption Song," I recognized a kinship and a truth. I thought about my own unspoken struggles with a sense of inferiority because of how dark my skin was. I'd always known that those thoughts were racist and morally abhorrent, of course: that was obvious on a conscious level. Still, I had truly thought that that stuff had rolled off my back, and I was outwardly more than confident. I felt unscathed.

Of course, it really is impossible to grow up in a world that despises people who look like you and not succumb to secret self-doubt at times. It quietly eats away at you, with a corrosive shame that is extremely difficult to extinguish because it goes unexpressed. This was especially true for someone like me, who was so devoted to being seen as cool and above it all. So "Redemption Song" moved me. And when Marley described how we were stolen from Africa to be placed into slavery in America in the song "Buffalo Soldier," it got me thinking about the heinous crime at the root of America's relationship with my people.

I felt like I wasn't alone for the first time; the sources of my pain had been named and were shared, after all. Moreover, undeniably brilliant and talented people had felt similarly. Even they were fighting the same demons, both from within and without. They had often themselves literally been hidden from view, like Ella Fitzgerald's voice appearing to emanate from the mouth of a white woman.

Gil Scott-Heron was another artist I discovered through Mark. His lyrics were intensely inspiring to me. I bought every album he'd ever put out and listened carefully to each song. When he skewered America's commercialism and the commodification and co-optation of rebellion in "The Revolution Will Not Be Televised," I felt like my world and experience were being expertly dissected and explained for the first time. The inanity of concerns like the soap operas that were always in the background back home was emphasized in lines like "women will not care if Dick finally got down with Jane on *Search for Tomorrow* because Black people will be in the street looking for a brighter day." The way television and commercial concerns about the right brand of products anesthetized us was not something I'd ever considered previously. My mind was opening.

From Scott-Heron I also learned about civil rights leaders like the NAACP's Roy Wilkins, mentioned somewhat disparagingly in another line of that song. And songs like "No Knock" taught me what I really should have already known about how unannounced police searches lead to abuse of power. It referenced the death of Black Panther Fred Hampton. Hampton was becoming prominent as an organizer in the 1960s, initiating free breakfast programs for children, arranging truces between major gangs, and leading unified actions against police brutality.

The J. Edgar Hoover–led FBI was so threatened by the Black Panthers and his leadership that they assassinated him, firing more than ninety bullets into his apartment while he lay in bed with his pregnant girlfriend. That no-knock raid occurred in 1969.[1] The FBI's racism and constitutional violations in the killing were so egregious that his family and that of another Panther who was also killed were ultimately awarded nearly $2 million. (The cost to the taxpayer of this and other regularly occurring examples of institutionalized racism is substantial.)

Listening to Scott-Heron's music, I felt that Mark and I weren't the only living black people who found materialism empty and longed for meaningful change. Here was a major recording artist, someone who got mainstream attention, not just some cat talking shit in the hood, who was saying what we all knew to be true. Here was a man who, Mark stressed, had a master's degree and had written a novel before turning twenty-one. This was not some random guy who just passed along street rumor, but a genuine scholar, someone who was highly educated and really knew history. That inspired me, and pushed me forward at times when I later thought about quitting school. And with brothers who dug Gil Scott-Heron, I felt like I'd finally found my people.

Back home, however, conditions were getting worse. The no-knock raids of the 1960s became even more prevalent over time: with the war on drugs as their rationale, by 2006, there were more than forty thousand military assaults on homes every year, with SWAT teams typically entering with no warning. Most of them occurred in black neighborhoods. In some of the tragic cases, police raided the wrong address and innocent people were killed.[2]

But, unfortunately, while I was just starting to understand a few things about black history and about who our enemies really were, I was also beginning to fall under the sway of some terribly misinformed ideas about drugs that were being spread for political reasons as a response to the so-called crack epidemic. I'd first become aware of the rise of cocaine during the home leave I took before I'd been sent to Japan.

I had received almost a hero's welcome when I returned home after completing basic training and what the air force calls "technical school."

My sisters were beaming, as proud of my achievement as I'd hoped they would be. I'd kept up with many of my high school girlfriends with those letters I'd written to keep my status up at mail call. I got to see all of them and hang with my friends. I was on top of the world.

It was Christmas 1984 and I was both glad to be home and glad I was not yet done with my travels and education. Just being away for such a short time had given me a new perspective on my neighborhood. But I wasn't yet able to accurately interpret how drugs like cocaine and the harsh drug policies that were starting to be adopted in its name affected my hometown. I did observe changes, however.

Although crack wasn't yet big in Miami, powder and freebase cocaine had already become quite popular by December 1984. As far back as July 1981, *Time* magazine had called cocaine "A Drug with Status and Menace" in a cover story illustrated by a martini glass full of sparkling powder. *Newsweek* connected cocaine to champagne, caviar, and other icons of wealth that same year. Eric Clapton's cover of J. J. Cale's "Cocaine" had been a major hit even earlier, back in 1977. Little gold or silver coke spoons had begun appearing around celebrities' necks in the late 1970s and early 1980s, along with winking (and some blatantly obvious) references in popular culture, especially *Saturday Night Live*, then at the peak of its popularity.

In the black community—as was true at the time for whites as well—cocaine had long been seen as a rich man's drug. But the price began to come down as the supply increased. This was especially true in Miami, which was a key shipping point where the drug came in from South America to be distributed to the rest of the country.

In the 1970s, marijuana had been the major Latin American illegal drug export to the United States. Miami was a major transshipment point. However, the use of the American military to interdict cannabis headed for America helped produce a shift to growing and selling the less bulky, more profitable, and easier to hide cocaine. Beginning in the late 1970s, the price of cocaine dramatically dropped for at least a decade, as the market became glutted.[3] The "rich man's" drug was about to become available to almost everyone. The South American marijuana trade began to collapse, but at the cost of creating a much more lucrative cocaine business.

I should explain here a bit of chemistry and pharmacology that is important to understanding the major distinctions between powder and crack cocaine, as well as many of the incorrect assumptions that have been made about these forms of cocaine and their effects. Powder cocaine is chemically known as cocaine hydrochloride. It is a neutral compound (known as a salt) made from the combination of an acid and a base, in this case, cocaine base.

This form of cocaine can be eaten, snorted, or dissolved in water and injected. Cocaine hydrochloride cannot be smoked, however, because it decomposes under the heat required to vaporize it. Smoking requires chemically removing the hydrochloride portion, which does not contribute to cocaine's effects

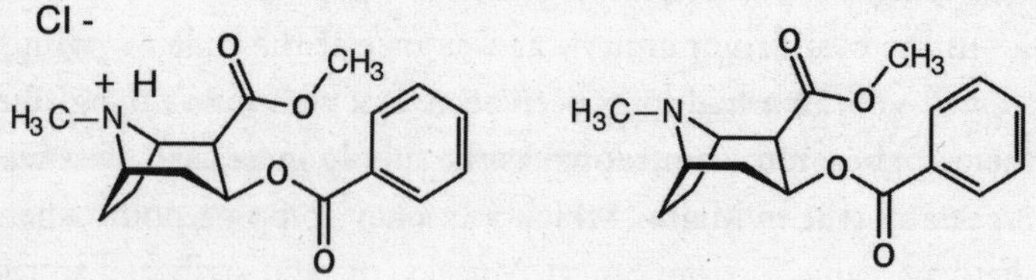

Figure 1. Chemical structure of cocaine hydrochloride
(powder cocaine), left, and cocaine base (crack).

anyway. The resulting compound is just the cocaine base (aka freebase or crack cocaine), which is smokable. The important point here is that powder and crack cocaine are qualitatively the same drug. Figure 1 shows the chemical structures of cocaine hydrochloride and cocaine base (crack). As you can see, the structures are nearly identical.

So, why do so many people believe that powder and crack cocaine are entirely different animals? This belief stems from a lack of knowledge of basic pharmacology, information that can help you to understand the effects of all drugs, not just cocaine.

Before a drug can affect mood and behavior, it must first gain access to the blood. From the blood, it then needs to reach the brain, where it can influence what you feel and the choices you make. An important basic principle here is that the faster a drug arrives into the brain, the more intense its effects will be.

Consequently, if we want to understand drug effects, it's essential to consider how the drug is taken, or in pharmacology-speak, the "route of administration." Route of administration is a key factor in determining the speed at which the drug enters the brain, and therefore, the intensity of the high.

Like most drugs, cocaine can be ingested in several ways. In the United States, it's rarely taken by mouth, although in some South American countries the oral route is common, usually in the form of chewing coca leaves, the plant from which cocaine is derived. Eating or swallowing a drug is convenient and tends to be safer because the stomach can be pumped in case of an overdose—this isn't possible with smoked or injection overdoses.

Once in the stomach, cocaine is dissolved and moved to the small intestine, where it can pass into the bloodstream. This process is called absorption, and many factors can influence it. If, for example, you've recently eaten a large meal, this will delay absorption and consequently, the onset of drug effects. By contrast, eating cocaine on an empty stomach speeds absorption and produces faster effects. As you've probably experienced, the same is true with alcohol. Drinking on an empty stomach produces more immediate effects than drinking immediately after having consumed a large meal.

After cocaine has successfully entered the bloodstream via the digestive system, however, its journey is not yet complete. Before it can reach the brain, it must first get through the liver due to the anatomy of the blood vessels it first enters upon leaving the intestines. Since the liver contains proteins that specialize in breaking down chemicals, including cocaine, in order to protect the brain and to make any poisons we eat less destructive, this can significantly reduce the impact of drugs taken orally.

This phenomenon is called *first-pass* metabolism. It's why—although many of them don't know the mechanism—experienced drug users seeking intense highs tend not to prefer eating their drugs or swallowing them as pills. The high from oral drugs typically comes on more slowly, larger doses are required to produce a strong sensation, and factors like recent meals and variance in liver processes can sometimes obliterate the effects entirely.

Snorting cocaine powder, on the other hand, bypasses the liver. Blood vessels lining the nose take it directly to the brain. Consequently, about five minutes after snorting a line, you "feel it." In contrast, oral administration takes half an hour to "hit you."

If you really want to get drugs to the brain fast, however, intravenous injection or smoking is the way to go. These routes produce the most intense high—and are associated with greater rates of addiction. Once injected, cocaine passes through the heart and is then transported immediately to the brain. As a result, the onset of psychoactive effects is almost instantaneous. That, of course, makes injection the riskiest form of drug use, not only because contaminated needles or unsterile technique can spread HIV and other diseases but also because overdose may occur quickly as well.

Smoking cocaine, on the other hand, avoids the risk of blood-borne disease but gets the drug to the brain just as rapidly as injecting. It exploits the large surface area of the lungs, which have lots of blood vessels to move the drug quickly from the blood to the brain, again, skipping the liver. Recall, however, that the hydrochloride form of cocaine cannot be smoked: those who sprinkle cocaine powder on their cigarettes or joints are probably wasting most of it because making powder hot enough to smoke tends to destroy it. In contrast, freebase or "crack" cocaine is stable at temperatures that vaporize it and that allows smoking rock to be as intense as injecting powder.

And that's why the two forms have the same addictive potential: cocaine powder can be injected, producing a high just as intense as smoking crack or freebase. Different intensities result from different routes of administration, but the drug itself remains the same. The following analogy illustrates this point. Consider leaving snowy New York for Miami Beach's South Beach via a luxury limo or private jet: both will get you to a highly pleasurable beach vacation eventually, but the jet will do so far faster. Similarly, injecting drugs intravenously or smoking them hits the brain more rapidly, producing a more immediate and intense effect than taking the drug by mouth. Despite this, the effects produced by the drug are qualitatively similar. Unfortunately, this is a distinction that lawmakers have yet to understand.

And to be fair, when crack cocaine first appeared, the truth was not initially clear. In the 1980s, even some researchers were not sure whether it was a new drug, which allowed hysteria and anecdotes to create a devilish image of it. The desire by casual cocaine users, those who snorted the drug, to distinguish themselves from people who shot or smoked the drug helped to fuel claims of unprecedented levels of addictive behavior caused by smoking crack.

The first media description of crack cocaine is believed to have appeared in the *Los Angeles Times* in late 1984.[4] At that same time across the country, 42 percent of arrestees in New York were already testing positive for some form of cocaine.[5] Nationally, 16 percent of all high school seniors reported having snorted cocaine at least once in 1984.[6] Statistics for Miami are not available for that year, but at least in my neighborhood, powder cocaine had definitely become a drug that some of my friends did on special occasions.

Indeed, smoking freebase cocaine made at home from powder had already become popular years before it started being brilliantly marketed as crack—often, unintentionally, through media scare stories that hyped the intensity of the high—as an entirely new drug. Only a few years earlier, Richard Pryor's infamous 1980 freebasing accident had drawn national attention to the practice of converting powder cocaine into smokable form. On June 9 of that year, the comedian was severely burned across half his body.

Initial reports claimed that he'd been set afire when a batch of freebase he was processing with the anesthetic gas ether exploded. That's quite plausible: ether is highly flammable and this manner of making cocaine base has great risk if people try to smoke near the ether. Then nearing the peak of his popularity, Pryor and his injuries became the subject of intensive media coverage.

As a result, freebasing instantly moved from being a fringe practice that few people in mainstream America had heard of to one that was seen as extremely dangerous. That helped prompt many freebasers to stop using ether and switch to the far less dangerous "baking soda" technique for making freebase. In this method, cocaine and baking soda are simply dissolved in water and heated until cocaine crystals form, making a distinctive cracking sound. No potentially explosive chemicals are involved. In fact, many believe that the "crack" made when cocaine crystallizes is the source of the name for the drug that is produced.

And so, crack cocaine began being sold as a ready-made product when dealers realized they could industrialize the freebase production process using this safe and easy baking soda method. The cheaper prices caused by the cocaine glut probably led to experimentation with new products and marketing ideas; the Pryor incident conveniently also raised awareness of the danger of the ether method. Crack may have been the ultimate result. My four years in the air force—from 1984 to 1988—coincided with the introduction and rapid spread of crack cocaine across the country. My home leaves during those years gave me snapshots of how the drug affected my neighborhood, although I first seriously misinterpreted what I saw.

During my first leave in 1984, I started to hear more and more about freebase. The first time I'd ever heard people talk about it had probably been when I was in high school. There was a set of twins who lived in the projects near me; I didn't know them well but I'd occasionally smoke reefer with them. Getting high one time, they told me to steer clear of freebase. "It's too good, man," one said. "Yeah, you can snort it but don't smoke it," his brother concurred. "That shit's not for rookies; it's just too powerful."

At that time, in line with my desire to always be in control, I had no interest. I didn't like the idea of not being able to stop doing something. The notion of an experience that overwhelming didn't sound at all attractive to someone who placed the emphasis I did on self-control. I wasn't even slightly curious. Back then, though—other than what I'd heard about Richard Pryor—I didn't see anyone I knew suffering serious negative consequences from cocaine. The guns and the risk of violence related to having a beef with someone were the same as they'd always been. That wasn't new.

So, cocaine use was definitely becoming popular by the time of my 1984 Christmas visit, and I heard some talk about it that year. There were rumors about a guy named Ronnie, who had always been known in the neighborhood for having the nicest ride. It was a Monte Carlo, sky blue, with a crystalized paint job that reflected the light just right. He had Trues and Vogues, which were the most coveted rims and tires. Ronnie put everything he had into that car; to say he loved it would be an understatement. Everyone who knew Ronnie knew about his car.

But the story was now that his ride was gone: "in the pipe," they said. The ride went in the pipe. Ronnie'd started smoking base and he'd stopped caring: that was the narrative. The Monte Carlo had gone up in smoke, along with his job and virtually everything else that had once defined him. "That shit's too good, man," was the way people phrased it. Ronnie's story supported the idea that smoking cocaine took you down, an idea I took on board with little critical thought.

Indeed, even though I smoked reefer myself, it never even occurred to me to question the military's drug-testing policy. Sure, I worried about being caught and I tried to minimize the potential consequences I'd face if it happened to me, but I accepted the idea that illegal drugs were bad and thought that expelling people from the service for using them was appropriate.

I alternately got high with Keith and his homeboys and discoursed with Mark about black consciousness. I took classes and began taking them seriously—but also stole movies from Gate 2 Street every week. My behavior was in transition: I was not quite yet a serious student, nor was I a complete fuckup. The balance could still shift in either direction.

In early 1986, I received word that Big Mama had suffered a stroke. She'd survived but wasn't expected to live long. The air force allowed compassionate leave in such situations. But at first I refused to take one: for some reason, I think, I couldn't bring myself to believe that her death was really imminent. I didn't want to even consider the idea.

I also had just six months left in my tour of duty in Japan, and I didn't want to fly twenty-four hours straight home only to have to do it again a few days later to return to a country that I hated. My first sergeant told me, "You're going to regret this."

He insisted that I'd be really unhappy if I didn't visit the woman who played such a big part in raising me, to say good-bye to her.

To ensure that I complied, he promised to arrange for me to be sent to my next assignment rather than back to Okinawa if I agreed to go home. And he kept his word. I flew back to Miami, wondering all the way if I was actually going to be able to see my grandmother alive. When I got there, Big Mama was barely holding on in the hospital. She couldn't speak and her face was all twisted up. She was in a terrible state.

Trying to protect me, my mother and sisters didn't let me get close to her: in my family, death was the business of women and they thought it would be too much for me to spend real time with her. I was at least able to pay my respects before she died. Moreover, the fact that she'd spared me another six months in Japan left me feeling pretty grateful. I was also happy to be back home.

Soon after she died, I heard from my commanding officer. He had good news: if I wanted, I could continue my service at home, in Miami at Homestead Air Force Base. Alternatively, I could go to England and start fresh in another foreign country. I felt inclined to stay.

Home was feeling comfortable to me again after a couple weeks; my girlfriends and friend-girls were welcoming and warm. After the lack of female companionship I'd suffered through in Japan, that was a relief and a joy. I felt nurtured and needed; I'd missed this so much. Why take the risk that another duty station might be as dissatisfying as Japan had been?

Since I hadn't spent any alone time with my father in a while, I went to find him. I wasn't looking for any particular guidance; I just hadn't visited with him yet. He'd always spent weekends drinking on the corner with his friends, so I went down to Seventy-Ninth Street and Twenty-Second Avenue and asked one of the guys if he had seen Carl Hart.

"Dunno, man," he said, coldly.

After having spent damn near twenty minutes asking several other people, I went back to the first one and said, "Yo, I'm his son Carl Jr."

Now his eyes lit up. Because of my military bearing and haircut, he hadn't recognized me. He'd thought I was Five-O, the police looking to harass my father. He directed me to Carl. After catching up a bit, I told him about my situation and my choice of assignments. I said I was leaning toward staying in Miami.

I talked about being there for my family and some other bullshit.

But my father wasn't having it. He looked me right in the eye, knowing well the real reason for my choice. I continued my story about responsibility and helping out after Big Mama's death. He stopped me. Carl didn't often give me advice, but he felt that he had to speak up now.

"Junior," he said, "pussy is everywhere."

He had instantly discerned my reason for wanting to stay. I was getting way too comfortable back home, possibly setting myself up to fail by being sucked back into the life I already knew, rather than moving on and at least trying something different. He knew all too well how easy it was to lose sight of your goals and drift aimlessly.

"You don't have to get it here," he said.

I just nodded. I didn't want him to know that he'd precisely pinpointed my motives. But over the next few days I thought about what he'd said and realized that he was right. The balance was back in favor of my success in college, which would truly begin in England.

CHAPTER 9

"Home Is Where the Hatred Is"

I came to the place of my birth, and cried, "The friends of my youth, where are they?" And an echo answered, "Where are they?"

—ANONYMOUS ARAB SAYING

S ir, we pulled you over because your taillight isn't working properly," the police officer said. He added, cordially, "We just wanted to let you know."

I had been driving into one of England's ubiquitous "roundabouts," which are similar to American traffic circles. I was on my second overseas assignment at Royal Air Force Base Fairford, in Gloucestershire, England. I was in my light green 1980 BMW 320; I'd purchased the car shortly after I arrived in the United Kingdom because I needed my own transportation to live off-base. It was around midnight on a summer or autumn evening in 1986 and I was on my way home from hanging out with friends to change into my uniform and work a night shift in the base computer room, where I was responsible for disseminating base supply reports. As always, it was drizzling.

The cops asked to see my license. While I was handing them the appropriate documents, one of them smelled alcohol on my breath.

"Have you been drinking, sir?" he asked, still respectful.

I admitted that I had had a pint, and complied as he administered a Breathalyzer test. I wasn't too worried that I'd fail: I knew that I wasn't intoxicated. Indeed, I blew well below the level that indicates any sort of impairment and the officers simply thanked me and let me go.

As I drove away, though, I suddenly realized that something was missing. I felt okay; my heart rate was pretty much normal. There was no dry mouth or sigh of relief. I'd just had an encounter with police that had involved very little tension or fear. It was peculiar.

The police hadn't flashed their lights at me; they hadn't stiffened or puffed themselves up when they saw that I was black. They'd been kind and respectful, not assuming that a black man in a nice car must be a drug dealer or some other sort of criminal. Even when they smelled alcohol on my breath, they did not become confrontational or judgmental and assume I was drunk. While my military ID might have helped, I'd still been treated like an ordinary person, not a second-class citizen or sketchy foreigner. I'd never had such an experience.

I thought back on a traffic incident I'd had with Florida police, which had also occurred late at night, in this case when I'd first returned home after boot camp in 1984. That had been completely different. Alex, my high school friend, had been driving his hideous brownish orange Pinto. I was in the front seat. The car—yes, it was the type that had been recalled for the minor problem of being at risk for exploding if rear-ended—was at least ten years old and probably looked twice that.

We'd pulled into a convenience store parking lot: in fact, it was that same old U'Tote'M that we'd frequented growing up. The shop was garishly lit, which usually meant it was open. Just after we'd stopped, Alex came around to my side of the car. He was carrying a large screwdriver, which was required to pry the dented door open so that I could get out. But we soon discovered that there was no reason to get out: the store was actually closed.

Just then, two cop cars pulled up and whooped their sirens at us, blinding us with their lights.

"What you boys doing here?" one of the officers drawled, full of undisguised contempt.

I produced my military ID, figuring that this might turn the situation around. After all, I was now part of the American security team, just like them, as I saw it. Alex simultaneously tried to explain about the problem with the car door. However, rather than placating the officers, this seemed only to antagonize them. Although I knew we'd committed no crime, I was flooded with apprehension. Everyone knew the many ways this situation could go terribly wrong. Images of police brutality flashed through my mind.

One cop said, "Where's your state ID; you know you're supposed to carry state ID." I wanted to say that military ID was a federally recognized form of identification and should be respected, but I could tell by this point that the best thing to do was to keep my mouth shut.

Meanwhile, the officers remained fixated on Alex's screwdriver. "What y'all doing around here?" they asked again. "You gettin' ready to open that door?" The implication was that we'd stopped at a store that we knew was closed in order to break in.

Fortunately, because they had nothing on us, they let us go after only a few minutes of disrespectful, condescending treatment. Then Alex laughed at my naïveté. He said, "You thought that military shit was going to help, air force boy. That shit don't work."

That same humiliating scene, which I and countless other brothers had been through before, would be poignantly described a few years later in Ice Cube's verses on N.W.A.'s 1988 "Fuck Tha Police." Cube's angry but brilliant analysis describes how the police routinely harass young black men mainly because of their race and gear, which may fit some stereotypical view of how drug dealers and criminals dress.

Driving home that night in England, I thought about how different things were there. My second foreign post had been an eye-opening experience, in more ways than one. Although I'd begun my college career in Japan—and had also had my first real exposure there to ideas about black consciousness and politics—it was in Great Britain that I really began to become knowledgeable about the profound effects of race in the United States and what it meant to be a black man from my background. I'd always known that shit was fucked-up, of course. But I hadn't had clear, precise language to describe it or to understand how best to fight back.

Having been schooled by Mark in Japan, I now schooled the younger brothers in England. And, as any good educator will tell you, convincing others of the superiority of your arguments is often the best way to master them and to fully convince yourself, too. In Great Britain, I used the social skills and leadership potential I'd developed during my youth to turn other guys on to Gil Scott-Heron and Bob Marley. I immersed myself in their music and studied their lyrics hermeneutically. They became my holy texts.

On the BBC, I watched documentaries like the PBS series *Eyes on the Prize*, learning more about the history of the civil rights movement and the real stories of the people behind the fight against segregation and other forms of discrimination. I also saw *Cry Freedom* and participated in actions opposing financial investments in South Africa, to help bring down apartheid. I began to regret having missed the activism and consciousness-raising of the 1960s and early '70s.

Ironically, as I began lamenting having been born too late to join the Black Panthers or protest the Vietnam War, I was

Getting ready to go out and party
in England while in the air force.

unaware that a new assault on black people was being launched back home. That was Ronald Reagan's war on drugs.

In 1986, in the United States there were isolated protests against Reagan—and in the United Kingdom, a much more visible revolt against Conservative prime minister Margaret Thatcher—but it all seemed pale in comparison with what I'd missed during the black power years. I didn't realize what was going wrong at the time in the States.

But being in England did give me a vital distance from which to analyze America. Though Britain was no prejudice-free paradise, its obsession with class and its early abolition of the slave trade made its racial politics different from ours. I wasn't constantly facing people who dismissed me before they'd even talked to me there. And English white women certainly didn't view black men the way American whites did in Miami. In fact, American military personnel—including blacks—were seen as having good jobs and greater opportunities than were available to the British working class. Our economic prospects were viewed positively, which was far from the case in South Florida.

Back home, one of the most conspicuous forms of racism I'd observed was related to interracial romance, particularly between blacks and whites. So when I started to date Anne, a tall, soft-featured brunette whom I met about three months after I arrived in England, I was especially aware of our respective races. As a boy, I'd always had to hide the brief encounters I'd had with white girls in high school and junior high. It was clear to me that seeing them publicly would bring nothing but trouble, so I stayed away. If I'd been on the street or in a store with a white girl in Miami, we would have run a gauntlet of stares and muttered remarks or worse. However, in London and even in smaller British towns, no one seemed to care. I moved in with Anne not long after we met.

And although she felt she had to diligently work to prepare me before she thought I'd be ready to meet her parents, her concerns about how they'd see me were about class markers, not race. Anne came from the British upper middle class. She was seen to some extent as the family fuckup because she didn't go to university. Her father was an aviator for the sultan of Oman and her parents spent most of their time in that country.

But as an American airman, I was seen as a "good catch" because of the economic opportunities open to me through the military and by virtue of being an American citizen. Compared to the Brits she'd dated previously, I was a definite improvement. Her parents didn't even object when I moved in with her to the family home. They had a huge four-bedroom house in Wootton Bassett, a suburb of Swindon; it was where I'd been headed when I was pulled over by the police that night. To assuage their slight discomfort about us "living in sin," I paid rent.

Before Anne introduced me to her parents, she carefully taught me to use silverware correctly and other table manners, which to me had been previously obscure. I didn't find this condescending or inappropriate. Instead, it was educational. I had a spongelike attitude and was determined to soak up any kind of potentially useful knowledge. I wasn't intimidated by the British class system because, even with all I knew about America's deep flaws, I still retained the notion of our country's ultimate superiority.

I learned a great deal from Anne and from observing British attitudes. The way they viewed American ideas about race, their support for civil rights and the equality of black people in the United States, confirmed for me that such positions were normal; this was how all thoughtful people should think about these questions. Fighting for civil rights wasn't some kind of "special pleading," or refusal to let go of "ancient history," the way it was often presented by white folks back home. Of course, criticizing the United States was easy for the Brits because it was another country; they weren't looking at their own issues. And their tolerance was far from perfect: they still had police brutality targeted at ethnic minorities and there was a persistent stereotype of Jamaican blacks as "lazy." Nonetheless, it was an improvement for me.

And hearing Gil Scott-Heron perform in a small club, with a mixed-race audience of about fifty people, further provided me with a real sense of being part of a conscious community. We all sat on the floor and he interacted and conversed with us, as though it were an intimate party and we were part of the music, not just an audience. Anne and I listened together. Times like that—and turning other guys on to him myself—energized me to take action and learn more.

Importantly, in England, I began to be repeatedly encouraged both by the professors with whom I formally studied and by the men I schooled about the black experience. They thought I had something special and that I could and should use my brain to help others. My job on base was to be a supply clerk in stock control, ordering necessary items via a very primitive computer. From salt for the runways to uniforms for the basketball team, if it had to be obtained and supplied, we had to order it, sometimes millions of dollars' worth at a time. But usually, this wasn't a particularly demanding position. There was plenty of time to think and study. Inspired by Scott-Heron and by my earlier talks with Mark in Japan, I decided that I'd become a counselor and work toward a career helping at-risk youth.

I worked a second job as an attendant at the gym on base, and played for the air force basketball team with games every Friday night and Saturday morning and practice daily after work. I took six to nine credits from the University of Maryland, which offered the courses on base, per semester. I also played on two British basketball teams: the Swindon Rackers and Swindon Bullets. My life was highly structured and all of this kept me extremely tired most of the time.

Nonetheless, professors began to take notice of my mind. Their reinforcement helped encourage me further. Not only was I being inspired by my teachers; I was also showing both them and myself that I could contribute academically.

Taking the required literature courses, I began to understand poetry and to see the hidden meaning in the allusions and references that had previously been obscured for me by dated language and unusual words. I read Auden and Shakespeare, and dove into the works of Gwendolyn Brooks, Claude McKay, Langston Hughes, and Sterling Brown. It was thrilling to be able to comprehend, value, and, most important, analyze for myself what intellectuals did. I felt pride in being seen as smart and capable by people who had been immersed in academia. It was as though I'd broken some kind of secret code and could now enter a world that I'd never even really known existed before. When I wasn't exhausted, I felt exhilarated.

England was where I started not just enrolling in college courses, but enjoying them; where I wasn't just doing the work because it had to be done but because I liked learning and wanted to know more and was good at it. I'd had brief moments like that in math as a child. I'd caught a few more glimpses of this possibility in Japan. But all of that was nothing compared to my ability in Great Britain to completely immerse myself in schoolwork. My instructors began to see a spark in me and this gave me more and more motivation, building my confidence.

I was still profoundly ignorant about the mainstream world, however. I still didn't know anything about the multitude of careers that mathematical talent could open up for me. I'd probably never met a scientist or a statistician or a mathematician. I'd no idea how heavily science relies on math and I couldn't yet conceive of myself pursuing a career as any type of scholar.

In fact, I was so lacking in the mainstream form of what academics call "cultural capital"—the kind accumulated in the United States by growing up in the white middle or upper class— that I made some mistakes that I now cringe to think about. Cultural capital is the knowledge of the way a culture—whether it is the culture of an institution, the culture of a country or community, or the culture of a social class—really operates. It's knowing the things that "everyone knows" in that class or place and the things that everyone automatically assumes that other people know.

For example, in my neighborhood back home, I was extremely high in cultural capital. There, people with cultural capital knew which employers were most likely to hire black people, where to get the best deals on food and clothing, which neighborhoods were "ours" and which were not, as well as who ran numbers and who had the best hookups for stolen goods. I knew the things that people of high status there should know, the things that kept me on top.

But in a middle-class neighborhood, cultural capital tends to include knowing things like which colleges are in the Ivy League and why that matters, in addition to the specifics of who has status, who does drugs, and where the best stores and restaurants are. The lack of relevant cultural capital is one of the things that maintains the sharp separation between people living in entrenched poverty and the mainstream. For instance, it allows some shady for-profit colleges and "institutes," which don't offer respected degrees—and sometimes don't even teach needed skills—to prey on the poor. When I was in Japan, in fact, I almost enrolled in one such "distance learning" program (these are conducted as online programs today), which was eventually shut down. Poor people often don't have the type of cultural capital that would let them know that these schools are seen as less than reputable by employers and by those who do have such cultural info.

Here's how little I knew about academia before I began my career. One of the concentrations offered by the University of Maryland on air force bases in Europe was "women's studies." I figured I was a natural for that. I certainly wanted to understand women and had spent much of my life trying to figure out how to get them to do what I wanted. While I might have gotten quite an education if I had ended up studying Angela Davis, bell hooks, Toni Morrison, and Gloria Steinem, my idea of "women's studies" and their ideas were not exactly similar. I had never even heard of feminism, let alone the black variant called womanism.

Although I can look back on this now and laugh, the results of lack of mainstream social capital are often not nearly as benign. Ignorance feels shameful; attempts to hide it can prevent learning and perpetuate the problem. When you publicly illustrate that you don't know what "everyone" knows, it can be intensely embarrassing. Many of the difficulties faced by people who try to move from the hood into the mainstream involve the lack of these types of knowledge, which marks them as outsiders and can lead to repeated humiliating experiences.

Though I ultimately discovered that women's studies did not contain the type of information I'd sought before I took classes in it, I was still naive enough to believe that psychology, instead, might hold the secret to understanding and manipulating women. The Psych 101 class I took consisted largely of Freudian ideas and I thought it was amazing that people could get paid to think up ideas about our minds and behavior like that. I figured

I could do it just as well. I decided that I would study psychology and it would be useful for both my potential career working with black youth and for my personal life. My relationship with Anne, my classes, and the air force itself helped me start to accumulate mainstream cultural capital.

Indeed, one of my professors, a black woman named Shirley Bacote, soon taught me something very practical that helped change my life. Like many of the black airmen with backgrounds like mine, I sent money home to my family, whenever I could. It was expected, even obligatory. From the outside, doing this looks commendable and altruistic: helping the folks back home who don't have the opportunities that you do.

But it can also be a trap, keeping you from investing in your own future. Shirley lectured about how black people don't trust themselves enough to really invest in themselves. She wasn't speaking directly to me when she said these things; she was teaching a small sociology course on race and class in America that had enrolled only one black man and a few sisters. However, I took her words to heart. I know that she must have known that most of us felt obligated to do this.

She explained that, while it was important to help your family and others in need, spending on your own education must come first. At school, you know you are developing marketable skills and that the money you spend there is working toward creating a better future. Back home, there will always be ongoing need. Invest in yourselves, she advised: that's the most sound way to invest in your family in the long run. Because unless you do that, you won't be able to advance enough to truly have the security to help effectively.

That stuck with me. I'd been contributing to supporting my family since I was twelve and started getting paid under the table. It had always bothered me, but I hadn't been able to put my finger on exactly why. I knew that my teenage jobs hadn't been like middle-class kids' summer jobs, to provide a little spending money for the kid and maybe a lesson in responsibility. Instead, I actually helped put food on the table.

If my sisters and I hadn't worked, there would have been many times when we would have had little in our kitchen cabinets or refrigerator. Without our childhood jobs, a difficult situation would have been even more difficult. It had never occurred to me that this wasn't the way family life was supposed to work. Parents were supposed to support their children, financially and emotionally, not the other way around, at least during childhood. I hadn't recognized how profoundly poverty and race had shaped my life until I'd left the country. I could see the way racism marred America much more clearly now.

For me, home was indeed where the hatred was, not just literally but in all the ways Gil Scott-Heron implied symbolically in his lament written from the perspective of a black heroin addict. The man in the song "Home Is Where the Hatred Is" is trying without success to use drugs to relieve his pain, pain so severe that he's considering never going home again. Listening to this song, I began to understand why someone might be driven to that sort of escape; to empathize in ways that I had not been able to do when I'd smoked marijuana and found the alteration in consciousness disorienting rather than liberating.

I still held conventional views about drugs as destroyers of life, however—and I would continue for years to come to buy into the idea that crack cocaine was the main thing that was devastating my neighborhood and other black communities back home. But I was also beginning to be able to see from multiple perspectives and to recognize more complexity than I'd earlier admitted. In this light, Gil Scott-Heron's own later problems with cocaine were only more tragic for me.

The public perception of his personal relationship with drugs and his songs about them unfortunately perpetuated myths about certain types of drug use. Because his own use appeared to be so pathological—and it seemed to have such a visible negative impact on him later in life—it played right into the stereotypes that use always leads to devastating addiction and that drug use by black people is the real source of our problems. Many of his antidrug songs recapitulated this conventional wisdom, without the penetrating analysis he usually brought to political issues.

While I listened then, though, I didn't yet recognize this. I saw drugs as being in opposition to black consciousness, as an obstacle. Fighting against drugs, listening to Scott-Heron's anti-drug songs, and sharing them with others was a way to fight oppression. It was a way to show that you were righteous. I didn't see then that the way we fought drugs actually made our oppression worse. I viewed the drugs as the problem, not our ideology around them or our treatment and law enforcement policies.

In fact, when I was home on leave in 1987, I became utterly convinced that crack cocaine was the cause of everything that I now saw as wrong with the neighborhood. I didn't realize it then, but I had reframed in my mind many things I'd seen around me. At the time, I was making the same mistakes in thinking that our leaders were. For example, I began to think that violence, the presence of guns in the hood, and the willingness of people I knew to use them were all caused by drugs. I left out the pieces—like my own family's experiences with domestic violence and parental absence and my personal experiences with gun crimes—that didn't fit.

I'd always looked up to my brothers-in-law and the rest of the older guys in our DJ group, seeing them as the baddest brothers in the world. But when I was back home, I started to hear them decry "these kids today." They said crack was turning nice girls into "chicken-head hos" and driving ordinary boys to "thugs, ready to cap a muthafucka." They couldn't stop talking about the lawlessness of the younger brothers coming up.

Of course, they'd schooled me themselves in all the nuances of respect and disrespect when I was younger. They'd trained me in the southern culture of honor that doesn't allow even the slightest dis, like a stepped-on shoe or dirty look, to go unchallenged. It wasn't like we hadn't ourselves carried guns and, in some cases, even used them to avenge incidents that outsiders would surely have characterized as trivial or even crazy.

Indeed, in the early 1980s, one of my brothers-in-law himself had been arrested after his bright-colored vehicle was used in a shooting: two people had been killed in the drive-by incident. No one was convicted for the crime because the shooter remains unidentified—but the chain of events that led up to the killings had begun when someone stepped on someone else's shoes. No drugs were involved.

The motives of young men who engage in these types of potentially fatal interactions over slights to honor are frequently portrayed as irrational overreactions. But these types of altercations that seem to have such petty origins are by far the leading motivation for deadly violence—contributing to significantly more crimes than the pharmacological effects of drugs. In their influential study of homicides in Detroit, Martin Daly and Margo Wilson concluded that the young men involved, far from being irrational, "may be acting as shrewd calculators of the probable costs and benefits of alternative courses of actions."[1]

This is how the calculation might be made. Before taking action to avenge a slight to honor, there are risks including the loss of reputation and status from being labeled a coward that must be considered. Conversely, possible benefits include impressing women and other men, resulting in an increased chance at both long-term survival and successful reproduction.

Potential costs of acts of revenge themselves, of course, include death, injury, or prison. But Daly and Wilson found that only about 10 percent of those involved who survived were ultimately convicted of a more serious crime than manslaughter, because the courts recognized that they were acting in self-defense. This means that they tended to serve little prison time. As a result, we can't conclude that such people are acting without considering the consequences: many of the risks were visible to them. And we can also observe that such crimes take place overwhelmingly among young men who have little to lose, with few resources and limited future prospects. This type of behavior had characterized male youth in my neighborhood long before crack cocaine was even invented.

Now, though, my brothers-in-law and the rest of the Bionic DJs claimed today's young men were different. It was all about crack. These kids had no code at all: "They'd just as soon smoke you as look at you. Shit done changed," they were saying. According to the oldheads, the "new" cocaine business meant that young bloods no longer followed any rules about respect. From hearing all this, I began to believe that crack really had changed things. The hot sound of rap, now ubiquitous, with its conflicted, ambiguous relationship to drugs—often glorifying slingers and hustlers, sometimes claiming to simply report what was "real," other times trying to scare brothers straight—also made it all feel new.

One night during my leave, I was driving through the hood with my brother Gary. At a stop sign, we were bumped by the car behind us. Shit, I thought, this is it; we're getting jacked. I'd heard about incidents like this where guys had simply been shot, point-blank, when they jumped out to look at the damage. What if these were some drug boys thinking that we were infringing on their turf? Or some jackers who thought that we were high rollers slipping (meaning, being less than vigilant given our status)? Or maybe Gary had done some shit I didn't know about and we were about to be murdered? I couldn't keep those images of kids who would smoke you for no reason out of my mind.

Gary, who probably had a gun himself, flew out of the car first to try to preempt any trouble. He soon came back laughing: the car behind us had been driven by a young woman. She and her friends thought we were ballers—pro athletes or other folks who had done well, now back in the hood—probably because we were driving Pop's brand-new Deuce-and-a-Quarter, a Buick Electra 225. They just wanted to flirt with us; nothing sinister. With my heart still pounding out of control, I checked them out. Gary got one of the girls' digits. But I just wasn't feeling it.

I felt like my neighborhood was becoming increasingly threatening; I saw constant news coverage of the "crack epidemic" and how it was destroying everything it touched. From the news, it seemed as though the senseless killings were spreading and unstoppable. In 1986, *Time* and *Newsweek* had each run five crack-related cover stories; the national media screamed out more than a thousand stories on the "scourge" in that year alone. Ronald and Nancy Reagan took to national TV to promote "outspoken intolerance" against drugs, calling them a "cancer" and asking Americans to join their antidrug "crusade."

I couldn't see it at the time, but what had really changed in my world was not the creation of an unprecedented wave of drug-induced violence and a codeless new group of predatory youth. It was how our problems were being described and explained. In the case of the media, politicians seeking reelection—of both parties—had spread the word that drugs were the cause of inner-city problems and that making war on them would fix things.

News organizations picked up the simple narrative, not questioning its assumptions.

In the case of my brothers-in-law, their reframing also had to do with having grown up. They'd settled down with jobs, mortgages, and kids. They no longer focused exclusively on their status on the street. All of these things—work, marriage, children—are strong alternative reinforcers. They aren't as available or attractive during adolescence to youth, but they become more so in early adulthood, as the view of what's appropriate and acceptable for one's age changes.

And once these alternative reinforcers began to matter more to my brothers-in-law, they started to see from a more mature and sophisticated perspective small incidents they once would have taken as challenges to their honor. These slights were no longer magnified by adolescence. More important, their jobs and families gave them other ways to see themselves as masculine that didn't require defending against every insult. Having children and jobs, of course, also meant that they had much more to lose.

The young guys really weren't any more lawless than we were; we had actually responded in exactly the same ways when we were their age. Some of the cues and certainly the fashions and music were different. But drug use was actually falling: in 1979, 54.2 percent of high school seniors reported using an illicit drug within the last year; by 1986, this had dropped to 44.3 percent.[2]

The same was true for murder rates. In 1980, there were 10.2 homicides per 100,000 people in the United States population; by 1986, this number had declined to 8.6. What's more, on September 25, 1986, the *Los Angeles Times* printed an article summarizing findings from a U.S. Drug Enforcement Administration report on crack cocaine. The report stated that media attention "has been a distortion of the public perception of the extent of crack use as compared to the use of other drugs." The DEA report further noted that crack cocaine wasn't even available in most places outside of New York and Los Angeles—crack problems and the later increase in dealing-related murders followed the wave of media hype about them, rather than preceding them. In other words, scare stories about an "instantly addictive" and violence-provoking drug served to spread crack cocaine, not accurately describe its use in most of America.

The effect of crack, when it had one, was mainly to exacerbate the problems that I'd seen in my home and in the hood since the 1970s. It didn't create the world of hustlers, dealers, and addicts celebrated by rappers or the underground economy that I'd always known. It was just a marketing innovation that added a new product to the drug world. The drug's pharmacology didn't produce excess violence. However, whenever a new illicit source of profit is introduced, violence increases to define and retain sales territory, then declines once turf has been marked out and the market stabilized. It happened in Miami first with powder cocaine and then again with crack cocaine, and the same pattern has been seen in numerous other locations with many different types of drugs.

But contrary to the image presented by hip-hop of immense wealth for virtually anyone who got into the game, the reality was that most dealers made about the same amount of money they would have made if they'd worked at McDonald's. Sociologist Sudhir Venkatesh meticulously documented the economics of crack dealing in his study of a Chicago street gang.[3] Spending several years on the streets with the group, he was able to gain the confidence of both the leadership and lower-level members and learn exactly what each person earned and how the profits were distributed.

While the risks of crack selling may not, on the surface,

seem worth the low salary ultimately earned, to many young men it seemed the best of a bad lot. At fast-food chains or in similar low-level jobs, these youth would have to wear dorky uniforms and submit themselves to often demeaning treatment from (typically) white bosses and customers, with rigid hours and little apparent chance for advancement. Selling crack, however, offered a choice of hours, the opportunity to work with friends, and visible routes to success, along with greater status among peers and potential girlfriends. The potential glory made the risk of prison and death seem worth taking.

But as with music or sports careers, dealing crack cocaine offered big cash only to a privileged few at the top of the pyramid. And the laws passed to "fight" the problem created an even crueler trap for those who succumbed to the drug's allure, whether users or dealers.

That's because, unfortunately, while crack cocaine itself wasn't an unprecedented phenomenon, what did genuinely change in the 1980s was the way the leaders of our community thought about the police and justice system. When I was coming up, we'd called the police "the Beast," and blacks had pretty much stood united in opposition to "crackdowns" on crime because we knew how unjustly they'd be enforced. But with the arrival of crack cocaine, black people themselves started to call for more officers and longer prison sentences, seeing the drug as turning their own sons and daughters into monsters who were beyond help.

The ongoing media emphasis on extreme pathological behaviors of crack users led people to believe incredible stories about it. For example, one of the most widely reported misconceptions about crack cocaine was that after one hit a person could become addicted. Addressing this issue at the time, Yale University psychiatry professor Dr. Frank Gawin told *Newsweek*,

"The best way to reduce demand would be to have God redesign the human brain to change the way cocaine reacts with certain neurons."[4] This is simply hyperbole. Even at the peak widespread use, only 10–20 percent of crack cocaine users became addicted. Another persistent stereotype was that most crack cocaine users are impulsive, focused only on getting another hit of the drug. Evidence from my own research (as well as from other researchers) shows that this too is incorrect. During my studies, I impose demanding schedules on crack cocaine users; they are required to do considerable planning, inhibit behaviors (for example, drug use) that may interfere with meeting study schedule requirements, and delay immediate gratification. Most meet these demands with no problems.

But the shift to a "law and order" perspective was real. Those who had once stood in opposition to the "get tough on crime" brigade, who had previously called for rehabilitation and community service, were now united with those who wanted more prisons and less mercy. In calling for the passage of the Anti–Drug Abuse Act of 1986—which ultimately created penalties for crack cocaine harsher than any other drug—Democrats and Republicans in Congress were equally rabid. They were eagerly trying to outbid each other in creating the harshest ways to "crack down on crack."

Indeed, when college basketball star Len Bias died on June 19, 1986, the hysteria reached an even higher pitch. At first, the twenty-two-year-old was believed to have died from smoking crack cocaine, but it later turned out that he'd used powder. The six-foot-eight University of Maryland student with a sweet jump shot had been the Boston Celtics' number-one draft pick. He died while celebrating having been selected to be on the team that had just won the NBA championship. His death had an outsize impact because the then–Speaker of the House, Democrat

Tip O'Neill, was from the Boston area and a committed Celtics fan. Rev. Jesse Jackson, in his eulogy for Bias, said, "Our culture must reject drugs as a form of entertainment, recreation and escape. . . . We've lost more lives to dope than we did to the Ku Klux Klan rope."

The cocaine-attributed death of Don Rogers, the twenty-three-year-old Cleveland Browns defensive back, later that month[5] only made matters worse. The closely timed deaths of these two young athletes in their prime contributed to the public belief that cocaine's effects were dangerously unpredictable. But they weren't placed in the context of the millions who had used or were using the drug without such effects.

In my research, I've conducted nearly twenty studies in which I've given cocaine to participants without incident: while it can, in rare cases, exacerbate preexisting heart problems, its effects in doing so are comparable to those that occur when people engage in other vigorous activities, like intense exercise. With increasing doses you get predictable increases in physiological measures like heart rate and blood pressure. Nonetheless, without any congressional hearings or much thought about potential negative consequences, the ill-conceived 1986 legislation was hurriedly passed.

Recall now that crack and powder cocaine are actually identical pharmacologically. Consider, too, that only a few decades earlier, Congress had passed tough mandatory drug sentences and then repealed them when they failed to have the intended effects. Almost immediately, it also became clear that the enforcement of the laws was having a biased impact—not because it was racist in intent, but because of the way law enforcement actually works and the way crack itself is sold.

Here's why. It is obviously a lot easier to catch people selling drugs in open-air markets than it is when people sell behind closed doors. Also, the more transactions a dealer or user makes, the greater the likelihood of being caught and arrested simply because more transactions means more opportunity to be caught in the act. One of the keys to crack's success on the market was the selling of very small doses at a low price, something that obviously increases the number of transactions needed to make a profit for the dealer; and because the actual dose of cocaine contained in street crack is low, it might necessitate users to make several trips. Since it was a new product, street marketing was also important to generate sales.

Unlike powder cocaine, crack was sold in smaller doses, making it affordable even to people with little money. These folks are both more likely to buy and sell on the street and to engage in more frequent sales transactions. Crack cocaine increased the prevalence of both street markets and frequent transactions in many black communities. Law enforcement agencies placed considerable resources in black communities aimed at arresting both dealers and users. This combination of factors meant that creating disparate sentences for crack would inevitably—even without any racist intent—put more black people in prison for much longer terms. And so, in Los Angeles, for example—a city of nearly 4 million people—at the peak of the crack epidemic, not a single white person was arrested on federal crack cocaine charges, even though whites in the cities used and sold crack.

Nonetheless, one of the key leaders in the war on crack cocaine was New York's black congressman from Harlem, Charles Rangel. He was then chairman of the House Select Committee on Narcotics Abuse and Control. In 1985, he'd criticized the Reagan administration for its "turtle like speed" in cracking down on drugs.[6] In 1986, his was one of the loudest voices calling for the imposition of tough measures to fight crack cocaine.

Rather than considering what was happening in New York

under similarly harsh legislation that had not "solved the drug problem" but had resulted in mass incarceration of black and brown people, he enthusiastically supported the most draconian drug war policies. That included the 100:1 disparity in sentencing for crack and powder cocaine, respectively, first ensconced in federal sentencing in the 1986 law. Seventeen out of twenty-one members of the Congressional Black Caucus, of which Rangel is a founding member, sponsored this law.[7]

Under the 1986 provision, a person convicted of selling 5 grams of crack cocaine was required to serve a *minimum* sentence of five years in prison. To receive the same sentence for trafficking in powder cocaine, an individual needed to possess 500 grams of cocaine—100 times the crack cocaine amount. In practical terms, 5 grams of cocaine results in 100–200 doses, whereas 500 grams results in 10,000–20,000 doses. From a scientific or pharmacological perspective, the disparity wasn't justified: it didn't accurately reflect any real difference in harm related to the drug. And soon, the Anti–Drug Abuse Act of 1988 extended cocaine-related penalties to persons convicted of simple possession, even first-time offenders. Simple possession of any other illicit drug, including powder cocaine or heroin, by a first-time offender carried a *maximum* penalty of one year in prison.

Overwhelmingly, those incarcerated under the federal anti-crack laws were black: for example, in 1992, the figure was 91 percent and in 2006 it was 82 percent.[8] While the intent may not have been racist, the outcome—lack of outrage and failure to change course in response to the disproportionate number of black men who were convicted, imprisoned, and disenfranchised—certainly was. The result, in many black communities, was an unchecked disaster that reverberates even today.

And as the late 1980s turned into the early 1990s, I began to see what I then thought were the effects of crack cocaine on my own family and friends. My cousins Amp and Michael became my own family's example. On one visit home around this time, I discovered that they had been exiled from my aunt Weezy's house because of their cocaine use. The cousins I'd once looked up to, who'd instructed me on sex and manhood, had now been kicked out of their own mother's house.

Instead of getting their own place, however, they had begun living in a toolshed in her backyard. It was the same one, in fact, that we'd tried to hide behind unsuccessfully when we were caught trying to smoke our first cigarettes as boys. Amid the rakes and lawn mowers in this ten-by-ten shack, my cousins had their new sleeping quarters.

When I came to see them, the shed was squalid, filthy. It had no electricity or plumbing, of course. Where were the cool older cats I'd admired and hung out with? Could these be the same brothers I'd looked up to, the ones I'd trusted to get advice from when I'd had my first embarrassing sexual experience?

These days, Amp and Michael weren't working or taking care of families; they were stealing from their own mother in order to buy crack. They were once even caught trying to steal their mother's washer and dryer to sell them to buy drugs. The only way their behavior made sense to me was if it had been compelled by a drug. At the time, I didn't recognize the roles played by factors like their failure to graduate from high school and Anthony's chronic unemployment. I didn't think about how we'd all engaged in crime back home, even without using drugs. I didn't know how Michael had gone from having a wife and steady job as a truck driver to living in that shack at his mom's. I didn't think about the difference the military had made for me. All I could see that differentiated me from them was their drug use.

I tried to find them to talk sense into them on a later trip home. But they dodged my sanctimonious black ass. They weren't going to let themselves feel humiliated. They also knew that I had nothing to offer them but empty words. The "just say no" rhetoric of the time wasn't effective for adults who had limited employment options and had previously said yes. And that's all I really had to give them then.

For one of my friends, though, the consequences of our failure to recognize the real problems that were behind the "crack epidemic" were even worse. I knew that when I'd joined the air force, Melrose and a few other friends had started slinging rock (meaning, selling crack cocaine) on the corner. They'd boasted to me about how girls would do "anything" to get crack and bragged about all the money they were going to be making. I hadn't paid much mind at the time because I knew that for all their talk, they were still living at home with their mothers or in other equally nonaffluent situations. Obviously, they hadn't made any real money.

I thought that their dealing was virtually all talk, like so many of the capers we'd planned in high school but never really gone through with. We had always been just about to get some real loot, always been about to grab the riches and fame that we knew were just around the corner. My experience in England had made the futility and unlikeliness of success in these endeavors transparent to me, and it now seemed a bit sad, embarrassing even. I hadn't expected their small-time hustling to amount to anything, good or bad.

But apparently, Melrose had been selling cocaine, on the 3900 block of Southwest 28 Street in Carver Ranches, pretty regularly. He wasn't moving large quantities and was no one's idea of a kingpin. Of all my friends, he was never one I expected to be involved with violence: although he was incredibly physically fit and an imposing-looking specimen, he was a genuinely good-hearted person. As a child, he'd been sent to that "special" school where he'd gotten no education at all, but he was gentle and no real threat to anyone. On August 14, 1990, he'd spent hours celebrating his daughter Shantoya's first birthday with her. Then he went out to the corner.

The guys who decided to jack him—some small-time dealers from another neighborhood who had targeted the spot where he worked—didn't know he'd just come from a toddler's birthday party. They didn't know Melrose was as kind and loyal a person as you'd ever meet. They didn't know him at all. They drove up and pulled out their guns before Melrose and his boys on the street had any chance to react. They made everyone on the corner lie down on their bellies, stealing their drugs and cash. Then for reasons known only to the killers, they shot Melrose in the back of the head.

Within three minutes, the shooters were caught and arrested by the police. But medical aid didn't come nearly as quickly. No ambulance at all arrived to help Melrose. His friend Michael's mother, Annie, called 911 four times, trying to get someone to take him to the hospital. Michael's sister Jackie ran to a nearby firehouse, where firefighters stood with their arms crossed, not responding to her pleas for help.

Annie had covered Melrose with a blanket and was sitting with him as he lay dying on the street for nearly twenty minutes before the paramedics finally showed up. An angry crowd of more than a hundred people later marched to the firehouse, furious about the slow response. Authorities claimed the rescue workers weren't authorized to help until police arrived to ensure that the shooting was over. But the arrests had occurred within minutes—and there was no reason to believe that there were additional gunmen at large.

Derrick "Melrose" Brown left behind four fatherless children. We'll never know if he could have been saved by a faster emergency response.

Melrose never had a chance. There were many critical experiences and policies that had led him to that corner, starting with his dismal educational history and the lack of economic opportunities it presented. But at the time, I blamed it all on crack cocaine. If he hadn't been slinging, if drug trade rivals hadn't come for him, he would still be with us, I thought. Forgetting my own early experience seeing my sister shot for no good reason and the equally senseless deaths of my friend's brother and the white motorcyclist I saw shot in retaliation for his death, I became convinced that crack had made everyone go crazy. And I soon decided to get involved in research that I thought could help do something about it.

CHAPTER 10

The Maze

It is one thing to show a man that he is in an error and another to put him in possession of the truth.

—JOHN LOCKE

Everyone in the psychology department knew about the class: some students even created T-shirts reading, "I survived experimental psychology," which they wore proudly afterward. It was among the most demanding courses in the entire curriculum, one of those make-or-break requirements that tend to weed out the distracted, lazy, uncommitted, or otherwise challenged.

Still, we hadn't expected to face a human version of the radial arm maze. We had seen this eight-armed, circular contraption in the rat lab and read about it in our texts. None of the thirty-odd students was quite sure what to do as we found ourselves, on a beautiful sunny North Carolina day, in the center of a large unpainted wooden structure, the size of a half-court in basketball.

It was about the third week of what was essentially my senior year of college, 1990. I was at the Wilmington campus of the University of North Carolina. I had no idea that this class and my professor, Rob Hakan, were about to change the course of my life. All I knew was that I was keeping my eyes on the prize, which for me at the time was simply graduating with a degree in psychology. I still had a vague idea that I wanted to work with underprivileged black children. I didn't have a specific path to that job in mind, beyond finishing college. Although that goal was tantalizingly close, if I hadn't taken Rob's class, I don't think I would have gone on to become a scientist.

Experimental psychology was focused on research methods, and the maze exercise initially seemed irritating to me. It was not exactly a challenge to determine which of the arms did, in fact, have a jar of Skittles or M&M's at the end of it. I felt slightly insulted to be treated, quite literally, like a lab rat. However, because I knew and trusted Rob, I went with it, figuring that he must have an important point to make by putting the class through this exercise.

And indeed when I tried to write up the results afterward, I immediately understood the experiment's purpose. I had to go back to check the number of the arms in the maze, the markers like red and blue dots of paint that helped distinguish between the arms that contained rewards and those that were empty and other specifics that I hadn't realized were important at the time. I could see why these details mattered and the importance of observation and measurement in experiments.

And as the semester progressed, I similarly began to discover the order and purpose that underlay much of what at first had seemed pointless to me in psychology. There was a beauty to the structure of this science and there were formulas for understanding behavior. What had seemed like arcane requirements for research and petty concerns were revealed as important ways to avoid bias. They were necessary to control the conditions so that you could ensure that the variables you were studying were indeed linked to the outcome of interest and not just incidental, but causal. This was a way to look under the hood of human nature by stripping away some of the confusing complexities. And it was quantitative, mathematical, solid.

Most important, I was learning how to think and communicate like a scientist; discovering for myself the profound truth of Einstein's quip that "everything should be made as simple as possible, but not simpler." For Rob's class, we wrote up a new experiment every week. That meant loads of practice, just as I'd needed to succeed in basketball. And as in basketball, practice helped me understand and learn how to work within the rules. As I learned them, I got better and more confident. All the while, my behavior was constantly being reinforced with "attaboys" from Rob and on tests and papers, good grades.

While I was having this awakening, Rob saw that I was increasingly serious and encouraged my questions. He wasn't one of those dazzling or charismatic professors who wow students with their outsize personalities and intellect; instead, he was quiet and soft-spoken. But he was young and attractive and his creative and challenging exercises and enthusiasm about his subject made him appealing. He stood about six foot four, with sandy brown hair.

I started hanging out after class to talk with him, then playing basketball with him on the psych department's intramural team. He turned me on to musicians I'd never heard such as Joni Mitchell and Bob Dylan. It seems strange now to think that I hadn't heard the music of these icons before meeting Rob at age twenty-three. But in my narrow world there didn't seem to be space or need for white folk singers. Rob also introduced me to books like Hermann Hesse's *Steppenwolf*, which helped me feel a real connection to academia. I well understood the feeling of wolfishness and the sense that one doesn't belong in polite society described in the book. Like the main character, who sees himself as a "wolf of the steppes," as well as a man, I felt I had a dual nature, too.

At the time, I was living with a woman named Terri Howard, a slender, light-skinned sister with huge brown eyes that made her look as though she might be the female twin of the musician Prince. She was a business major and we would be together for four years. While I tried to look respectable and behave respectfully toward her pretentious Republican mom and her mom's new husband, they seemed to think that a man like me with three gold teeth and whose speech was still filled with heavy street vernacular was too rough around the edges for their Terri. I took considerable comfort in learning that a leading German intellectual who lived more than a century earlier than I did had struggled with many of these same issues.

Further, Rob made it clear that there was room in research for people like me, those who had not followed the traditional middle- and upper-class route to academia. Indeed, his lab team at the time was a collection of seeming misfits—all of whom went on to later success in medicine and research. One guy was a bona fide Deadhead, complete with long hair, beard, and hippie paraphernalia. Another started as a skinny, highly distracted dude who had so much energy that he smoked pot to calm down. His intensity made people nervous. There was also a highly driven married couple whom we called "the spouses" (their last name was Strauss), whose visible competitiveness stood out at laidback UNC-Wilmington.

After I got one of the highest grades in his class, Rob encouraged me to enroll in an advanced independent study course that he would supervise. It was then called advanced physiological psychology, but it would now be labeled as behavioral neuroscience. In order to complete the coursework, however, I had to learn some new skills. The first thing Rob wanted me to do was to learn to operate on the brains of rats. Although I was much more interested in helping him with survey research he was then conducting on human sexuality, he was out of funding for that project. He convinced me that if I learned how to do rat research, I just might help unlock the secrets of the human brain, cure addiction, or at the very least, make a career for myself in science. I was flattered by the attention and wanted more such praise. I wasn't sure at first, but over time I began to think I might be able to do it.

Much of my confidence came from the fact that Rob was very clear with me that hard work was what mattered most. Because he kept reinforcing that idea, I wasn't as intimidated by the subject matter and the actual brain surgery I had to do as I otherwise might have been.

"There's a place for people like you and me in science," Rob would say, meaning for those who weren't the obvious nerds and geeks, the ones whose persistence and diligence could allow them to overcome any educational deficits they had. My dismal high school education had left me without the science background and vocabulary expected of a researcher, but Rob saw that I was willing and able to do everything required to remedy the situation. I had already shown both him and myself that I wasn't afraid of hard work, even if it meant going back into the maze repeatedly.

I'd had to run a labyrinth of my own before I found Rob and the two other mentors who guided me into science. When I first got out of the air force, it was not at all clear where my future lay. After leaving the service in 1988, I'd first gone home to Miami. I was about thirty credits short of a college degree and planned to finish my coursework at Bethune-Cookman College (now

Bethune-Cookman University) in Daytona Beach. I'd saved up
a few dollars and was feeling pretty good about myself.

But having been in England and in the military, where I
had considerable responsibilities, being back in the American
South felt like stepping back in time. My old friends couldn't
even imagine having done many of the things I'd done in the air
force; their visions of the future were stunted by their lack of
education and inexperience with anything other than the small
neighborhood where they'd spent virtually all of their lives. I
could now see the limits of this point of view, rather than simply
accepting it as "how it is."

Another experience further reinforced my sense that there
had to be more than this for me. A few months after my discharge,
I interviewed at the then-nascent Rent-A-Center company for a
managerial job. The chain rents out furniture, computers, appliances, and other essentials to people with little money and/or
bad credit, charging high interest rates and offering hope of
eventual ownership if they can keep up the payments.

By this point, I'd become worried that I would run through
the money I'd saved up during my time in the service. I also
wanted to save even more to use toward finishing my degree.
The regional manager who interviewed me recognized my skills
and talent. Indeed, he almost immediately suggested that I work
at an existing store for a short time, just to get to know the trade
so I could be prepared to manage my own new location in a few
months.

But the day I started turned out to be my last day at Rent-
A-Center. The store was located in Carol City, on 183rd Street
and Twenty-Seventh Avenue, an area I knew well. Its customers were overwhelmingly black. And yet I was the only black
employee at the store. Worse, the local store manager treated me
with disdain. He asked me to do tasks requiring physical labor

and generally treated me like a short-term, dumb-ass minimum-wage drone, not a managerial candidate. He spoke down to the customers, making subtly patronizing remarks and refusing to play the radio station that broadcast the music that we liked. I quit at the end of the day. I simply couldn't take being treated like that anymore. I knew I deserved more respect. And I began to see that I wasn't going to get it working in my old neighborhood.

People like my cousin James and MH thought I'd gone crazy. From their perspective, I'd quit a good job for no reason. I didn't know how to explain it to them. I knew I couldn't engage them in the type of intellectual discussions about the books and song lyrics and poetry that had helped raise my consciousness while I was in the service. I didn't feel like I could reach them so I didn't even try. I now realize this must have seemed like I thought I was too good to do that type of work, but I didn't know how to bridge the expanding gap between us. I didn't even know how to parse that distance to myself.

One of the few people I connected with back home was Yvette Green, a former girlfriend who was then studying nursing. We'd go to that Denny's where I'd once "dined and dashed" with my high school friends—but now I was spending hours with her, reading and discussing literature. She gave me support, comfort, and peace of mind. Indeed, one of my greatest regrets in life was losing touch with her when I did leave Florida.

When I was home, though, I mostly just felt out of place. I had expected to slide easily back into that world, even to educate people and show them how cool I was by sharing with them the skills I'd learned for success. Instead I discovered repeatedly that I didn't know how to do that. My hometown itself began to seem increasingly foreign to me. In the air force, I'd unconsciously abandoned the habits of mind that had desensitized me to the daily wear of being condescended to and disrespected, but I didn't yet have a way to appropriately communicate my new perspective with those who still needed those defenses.

I found it more and more difficult to connect with my closest friends and family members. I wanted to discuss the larger societal issues that trapped so many people like us in those horrible conditions. But they were more concerned with immediate issues like how to pay this month's rent and how to put food on the table today for their kids. They had little interest or time for what someone called my "academic masturbations."

I wanted to work on changing the world and all they wanted was work. I didn't fit anywhere. It was like that awful time in adolescence when you feel half-formed, no longer a boy but far from being a man, as well. Everything felt somewhat awkward. I soon realized that I couldn't stay unless I wanted to relinquish the new self and changed vision of the future I'd constructed in the air force. To live at home without going crazy, I'd have to reembrace what I now saw as a very limiting worldview and pattern of behavior. I knew I had to resist that.

And as this conflict between my new self and my old ways increased, I got in touch with my cousin Betty. She had moved to Atlanta after her divorce was finalized. She invited me to stay with her there. I could take the credits I needed to complete my college degree at Georgia State University in Atlanta. Also in Atlanta was Patrick, my good friend and fellow airman with whom I'd served in England. He, too, had recently been discharged. He was one of the few people I knew who understood the transition I was facing after leaving the military.

Given my experiences at home, I figured that anywhere else would probably be an improvement. When I first arrived in Georgia, Betty had a house in Stone Mountain, just outside metro Atlanta. But money woes forced her to move to a smaller place in the same town. Unfortunately for me, however, Atlanta really wasn't much different from Miami. I didn't really find the move any more conducive to furthering my educational or personal goals. However, I did meet Melissa, the woman who turned me on to cocaine—and my relationship with her, ironically, is what led me to Wilmington and Rob Hakan's class.

My introduction to cocaine and Melissa actually started with a bad experience with marijuana. That incident not only was the beginning of my relationship with her but also gave me more insight into the effects of marijuana and into how environmental factors can strongly affect the drug experience. Additionally, it should have made me more skeptical about what I was hearing on the street about drug use and about what I'd later hear from addiction researchers, but I wasn't yet thinking critically enough to recognize this.

I met Melissa one summer morning in 1988 in the laundry room of the apartment complex where I lived with Betty. I was home at that time because I hadn't yet enrolled in school and was working some night shifts for UPS to make money before I went back. Melissa was a gorgeous caramel-skinned woman with long hair. She wore colored contact lenses that made her eyes look blue, an effect that I found disconcerting. Her aunt, who was also extremely attractive and about the same age, was also doing her wash when I met Melissa.

Over the course of our conversation, I discovered that the two women smoked weed—and as any self-respecting player knows, if you've got drugs, you can get girls. I said I had a hookup and invited Melissa to stop by Betty's place that evening to hang out. Then I called Patrick, who typically had at least some reefer on hand, and had him bring it over.

That afternoon I also watched *Oprah*. The show was then at the height of its popularity among wannabe in-the-know black people, so I was a daily viewer. The program that day featured a group of young, attractive women who were known as the "Rolex bandits." Their trick was to target men with Rolexes in bars or clubs and get them so drunk or high that the women had little difficulty pretending to seduce them and thereby steal their expensive watches. I wasn't paying all that much attention but got the gist.

Early that evening, Betty left to go out with her boyfriend. Melissa arrived not long afterward, unexpectedly accompanied by her aunt. I understood; she didn't yet know me and wanted to take her time. Visiting a man's house alone at night might set up unwarranted expectations.

After some small talk, the three of us passed around a joint. But although I'd continued to smoke reefer occasionally while in England, I had always stayed mindful of the fact that I could be urine-tested at any time. I generally didn't inhale much: both for that reason and because I still found some of its psychedelic effects uncomfortable and disorienting. Although I'd smoked some in Atlanta with Patrick, I didn't have much of a tolerance.

But wanting to seem cool to impress the woman I was attracted to, I smoked far more reefer than I'd intended that night. At first we had a good time, just laughing a lot, making silly jokes. After about an hour or so, however, I started to become paranoid. It started with a nagging sense of unease. And then I became convinced that these two suspiciously beautiful women I'd picked up were Rolex bandits like the ones I'd seen on *Oprah*.

Needless to say, I did not have a Rolex, nor was there anything of great value to steal in Betty's apartment. Melissa and her aunt did not behave in any way that was at all suspect. It was highly unlikely that the day I watched an *Oprah* episode about females who preyed on men using sexual enticements to rob them I would have such an experience myself.

Nonetheless, once the idea was planted in my head, I couldn't get rid of it. Everything seemed to be telling me that these ladies were up to no good. I tried to tell myself to chill out, but it was to no avail. The paranoia became almost unbearable. I had to do something. To everyone's surprise, without warning, I suddenly stood up and said, "Y'all gotta get the fuck out!"

What had been a pleasant evening suddenly turned strange. They both looked at me and said, "What?"

"You gotta go. Now," I said. There was a serious edge in my voice. They froze and then began hastily gathering their things to leave.

I certainly was attracted to Melissa and she seemed into me. But at that moment, I thought she was just trying to use me. I was so paranoid and insistent—and probably frightening—that the party stopped right there. I thought I'd never see her again.

As silly as that experience seems in retrospect, it illustrates some important issues about drug use that have critical implications for how we understand and deal with it. A drug's effects are determined not only by the dose and the way it's taken into the body but also by many different characteristics of the user and her or his environment.

LSD guru and onetime Harvard lecturer Timothy Leary first popularized the notion of set and setting as being crucial to the psychedelic experience. By set, he meant the mind-set of the person who has taken the drug: their preconceptions about the substance, expectations of its effects, and the person's mood and physiology. Setting encompasses the environment: the social, cultural, and physical place in which the drug use occurs. It turns out that these two factors affect all drug experiences, not just those with psychedelics. Though some of Leary's approaches had serious limitations, the concept of set and setting remains useful and they are crucial factors in understanding drug effects. The major point here is that psychoactive drug effects are not determined by pharmacology alone. It is the interaction between biology (the drug's effects on the brain) *and* environment that determines drug effects on human behavior. This is why attempts to characterize drug effects on human behavior by solely examining the brain after drug administration are inadequate and naive.

My set and setting the day I kicked Melissa and her aunt out of Betty's apartment weren't especially conducive to a "good high." That *Oprah* episode had raised the possibility in my mind that sexy women were likely to be predators and con artists—so my mind-set wasn't likely to make me feel comfortable getting high with women I didn't already know and trust. My reduced tolerance also increased the chances that I would experience paranoia from smoking more than I was used to handling. With THC, the primary active ingredient in marijuana, higher doses taken by inexperienced users increase the odds of negative side effects like paranoia or anxiety.

Set and setting can explain a lot about the variability of effects reported by users who take the same drug and about why different environments can produce different behavioral responses to drugs. The divergent responses of the Rat Park rats (see chapter 5) that eschewed morphine in favor of family and socializing with other rats, and the isolated rats who took dose after dose of the drug, are one example. Another one is the differing experiences of smoking cocaine found in Wall Street traders and among homeless cocaine smokers. The homeless people experience far more paranoia and fear than the executives do because wealthier users are more likely to be sheltered from frightening consequences like being arrested. The setting of the drug use can profoundly influence behaviors that are often attributed to the drugs themselves.

The night I got high with Melissa and her aunt, I couldn't

sleep at all. Now I know that adequate sleep is essential for the health and survival of an individual and that severe sleep loss, even without drug use, can cause hallucinations and paranoia. As a result, even the next day, when I tried to make a deposit at the bank, the paranoia was still with me. Standing in line, I felt as though the cameras were trained on me specifically. I got so freaked out that I left without depositing my UPS paycheck. But I realized early on that this was the result of smoking so much weed and simply waited for it to wear off.

And fortunately for me—and, as it turned out, my academic future—Melissa had a serious crush on me. Several days later, when I ran into her again, she came over immediately and asked me if I was all right. I laughed off the incident and before long, we were seeing each other. Melissa would become my girlfriend for the next year and a half.

About a month later, Melissa introduced me to cocaine. One of the local dealers was also pursuing her, though she wasn't much impressed by him. He asked if she indulged, seeing an opportunity to spend time with her; she said yes but often secretly stashed the cocaine he gave her so we could do it together. I wasn't especially interested in the drug. But when she took it out for the first time, I didn't think it would be cool to say no.

It was still 1988; at the time you couldn't turn on the TV or see a headline without being confronted with a story about the horrors of crack. I still knew nothing more than street lore about drugs, but even that far into the 1980s among the people I knew, powder cocaine had retained its glamorous associations with wealth, celebrity, and sex. Snorting it was perceived as fun rather than risky or particularly addictive. I didn't see any harm in trying it and thought that Melissa knew what she was doing, although I later learned that she wasn't actually an experienced user.

And sniffing my first line through a straw, I thought it was great. I had a sense of mastery and found relief regarding any anxieties that I might have been feeling about the evening. But while the drug made Melissa peppy and talkative, I found it calming and became more contemplative, perhaps because I also drank Schlitz malt liquor when I did lines. (Interestingly, although most drugs are not taken solo, little research focuses on the effects of drug combinations.)

Like many Gil Scott-Heron fans, I'd also taken to writing poetry. After a few lines, I loved nothing better than writing. As many cocaine fans discover, while the drug can produce stimulation and mental clarity, you may also come to see the most banal thought as being the height of brilliance. Under the influence of cocaine, dull or usual thoughts sometimes seem more important or significant than they are during nonintoxicated states. This is one of the main reasons that people take drugs: to alter their consciousness. And, as far as we can tell, humans have sought to alter their consciousness with psychoactive agents (often plant-based) ever since they have inhabited earth. There doesn't seem to be an end toward this pursuit anytime soon. In other words, there has never been a drug-free society and there probably will never be one. So slogans like "our aim is a drug-free generation" are nothing more than empty political rhetoric.

Nonetheless, although I thoroughly enjoyed it, I never developed the intense craving for or compulsive use of cocaine that some users describe. I knew that if I developed a serious cocaine habit it would have jeopardized my ability to earn money, which in turn would have jeopardized my housing arrangement with Betty. With no money or accommodations, I seriously doubt that Melissa would have remained interested in me. So, when cocaine was available— and Melissa and I did it about twice a month for a few months—I often did want more as we enjoyed our supply. When we ran out, however, I never found myself unfolding the packet to see if there was perhaps a hidden clump left or looking on the mirror for stray flakes. I didn't even consider going out and buying some. It was certainly pleasant and I definitely enjoyed the sense of clarity it gave me. But it wasn't irresistible to the point that I was willing to jeopardize the things—earnings from work, housing, and Melissa—that allowed me to indulge in the first place.

Yet again, I'd had the experience of most drug users, the not particularly exciting nonaddiction story that never gets told. I was in the 80–90 percent of cocaine users who do not develop problems with the drug, the group that rarely speaks out about their experiences because they have nothing much to say about them or because they are afraid of being vilified for having taken an illegal substance. In the current political climate, it is not surprising that many drug users do not speak out about their experiences. I served as an expert witness in multiple court cases where mothers have had their children removed from their custody simply because they admitted to smoking marijuana. My testimony on behalf of these mothers, explaining that it's inappropriate to conclude that someone has a drug problem simply because they admit to illegal drug use, didn't seem to matter. Since we tend to hear from that problematic 10–20 percent, their experience is incorrectly regarded as the norm.

Indeed, when I began researching drugs myself as a scientist, I first discounted my own personal experience as being aberrant, falling for the propaganda that continuously puts pathology at the center of the drug dialogue. I ignored my own story, just as I had when I didn't notice that problems in my neighborhood that were later attributed to crack cocaine had actually preceded it.

Because my ties to Atlanta weren't particularly strong, when Melissa suggested that I move with her to North Carolina and take a job at her mother's restaurant, I agreed. I became their short-order cook and manager. The idea was that their restaurant was going to be a huge success and make us all lots of money. Simultaneously, I enrolled at UNC-Wilmington in 1989, still set on finishing my degree. I managed to get some Pell Grants to cover the tuition. If that didn't work out, I figured the restaurant job would.

Without my relationship with Melissa, I might never have become a neuroscientist. If I hadn't met her, I wouldn't have moved to Wilmington and wouldn't have taken Rob Hakan's experimental psychology class at UNC-W. Also, I would have never met my two other crucial mentors at that school, Don Habibi and Jim Braye. I don't know if I would have completed my education without these three men. Due to my seemingly interminable work at the restaurant, I almost dropped out within a few months of starting.

Managing a restaurant and being a cook is hardly a part-time job. Pretty soon I found myself working twelve- to sixteen-hour days for minimum wage, dumping the grease into the grease pit when my shift finally ended at 1 a.m. and wondering how the hell I'd gotten there. I stank of sweat and cooking oil and every part of my body was tired. The long hours meant that I had little attention left for my classes and even less time to do homework. In my first semester, I barely managed to make Cs.

Without being aware of it, I began to slip away from academia. My air force goal of becoming a counselor to uplift black youth started to seem like a foolish pipe dream. I was called into the financial aid office because I was required to maintain a 2.0 to keep the Pell Grant funding. My grades were so low that I was in danger of losing it.

But during that same time, I also took a philosophy class with a young white professor named Don Habibi. It was his second semester teaching and he was the most intellectually curious person I'd ever met. It seemed like he knew something about everything—and yet he treated me like my perspective was unique and important as well. We connected. As a Jewish man who felt out of place in the South, he understood, I think, some of the alienation I felt, too.

Later, when I moved into the building where he lived, we became even closer and he encouraged me to continue to take the academic opportunities that began to present themselves. He was single and admired my ability to meet women; I respected his intellectual achievements. I'd take him to black clubs and in return, he taught me many essential aspects of cultural capital associated with growing up in the white middle class. When I first took his class, though, it was still not completely clear that I'd be able to stay in school.

Luckily, however, I had also found another mentor who refused to give up on me. Jim Braye was one of only three black men on campus who were in professional positions at the time. He did not teach, but rather worked in the administration as the director of career planning and placement. He was a retired army colonel with a rich, deep baritone voice that made him sound like Paul Robeson. My time in the air force had given me great respect for any black man who had moved up through the ranks in the military, particularly as early as he had, which was during the Korean War.

A friend of mine who had also been in the military had introduced us. I followed up with Jim and he had actually helped me enroll at UNC-W in the first place. As had happened many times before in my young life, chance placed an opportunity in front of me. I saw it and grabbed on, as if it were a life preserver.

Soon Jim began spending hours with me, teaching me new vocabulary and even how to pronounce words that I often stumbled on like *apocalypse*. He had a calendar with a "word of the day" to learn, and he'd drill me on them as the weeks went by. When he saw that my restaurant job was getting in the way of my education, he kept his eyes open for job openings in psychology for which he thought I'd be qualified. He had me do mock interviews in his office. He taught me about the hell that black men—even those with his accomplishments—catch in the white world.

Often, however, he'd just let me hang out and soak up his wisdom. I wasn't afraid to seem "dumb" or "uncool" in front of him because it was so clear that he knew much more than I did. Before long, he was like family. I could tell that he understood my struggles.

Sometimes when he saw me coming he would take one look at me and say, "Time for a shot in the arm." He could always tell when I needed a lift. Then he'd close the door to his office behind us and tell his secretary that we weren't to be interrupted. I loved listening to him because he sounded so authoritative and was so wise. He wouldn't let me get discouraged.

Most of the other students I knew didn't recognize what he had to offer because they hadn't been in the military. But I could see that he'd learned how to survive in a biased world and I paid attention. I wanted what he had and wanted to know exactly how he'd gotten it. It was because of Jim that I finally quit the job at Melissa's family restaurant and got an entry-level position that did not require a degree at a child psychiatric hospital, which had more student-friendly hours. And that was why, when I took that experimental psychology class with Rob Hakan in my last semester, I was ready to learn and be inspired.

My best friend and classmate, Walt, was a brother with whom I used to listen to the latest Public Enemy LPs. We'd sit for hours critiquing every lyric and relating them to our current situation at UNC-White (the name that black students called the university due to its low number of black students and faculty, despite being located in a town with a large black population). Walt couldn't understand why I would spend so much time with white men like Rob and Don. I had to explain that I needed support from those who'd forged the kind of career I wanted. No matter how different they seemed from us, they were actually more like us than their colleagues, I argued. Walt couldn't wrap his head around this thought.

Indeed, research shows that having a white male mentor is advantageous to women and minorities in science. When a field contains few members of historically excluded groups, having a mentor from the privileged majority can open doors. In one study of sociologists, for example, blacks with white male mentors were found to be more likely to be on track for tenure and to get a position at a major research university, which led to publications in better quality journals and greater academic productivity.[1] For me, both in college and graduate school, having a variety of mentors with different experiences and strengths made a massive difference. I was happy to receive all the knowledge and insight I could from wherever it was offered.

Of course, making good use of multiple mentors means recognizing their specific expertise: a white male mentor may give useful advice on science but be less knowledgeable or effective in advising on the race-related challenges a black student faces.

Even after I'd found my three mentors, however, I hadn't completely left my old life behind. Money was a constant issue. None of the jobs I had paid more than six dollars an hour and once I got involved with Rob, I was spending more and more time at the lab, which initially didn't pay at all. When Melissa and I broke up in November 1989, I needed to find a new place to live because she had paid half the rent. A woman who ran a record store that specialized in reggae allowed me to stay in her store for a short time, until she hooked me up with a Jamaican named Dwight, who wanted a housemate.

Dwight was a cool brother with long dreads that he wore covered with a hat. He was also a high-level marijuana dealer: he had operations in Miami and Brooklyn as well as Wilmington. I didn't care that he was in the game; his being a drug dealer was not my concern. I wasn't about getting into anyone else's business. I needed an affordable place to live and he had one. He knew that I knew but it wasn't something we talked about. Besides, his position within the drug game was high enough so that he himself never possessed marijuana. So I didn't have to worry about our place being raided by the police or robbed by rival dealers. He was a low-key, mellow guy who had also worked in construction. Well, he didn't actually work in construction; he just kept his union dues current to give the appearance of maintaining legitimate work.

About ten years older than I was, Dwight soon became impressed when he saw me studying and getting involved in lab work. He saw my vocabulary improve as I practiced. He soon thought I was some kind of brainiac and began bragging about me and my scientific future to his friends. Meanwhile, I was living way above my means, maxing out the multiple credit cards that were then being sent to college students as though the companies were giving away money. When the bills inevitably came due, I first pawned the saxophone that I'd once tried unsuccessfully to learn to play. Then I asked Dwight about getting in on the dealing action.

He flat out refused me. In the way that people in the life often look out for those who have alternatives, he didn't want me to get dragged down. He said it was ridiculous for me to even think about and that I was too smart for this. He did, however, begin letting me stash money for him. Sometimes I kept it in the room where we housed the rats for my research. I don't know if he even really needed me to do that or if he was just trying to give me a way to feel like a man who wasn't reliant on him for charity. Still, he helped me get through my crunch and was another man in my life who refused to let me give up on myself. (Dwight himself, sadly, was later shot to death in Brooklyn; I don't know the exact circumstances of his killing.) I slowly got out of debt and managed to stay that way. With Dwight's help, I managed to keep my nose to the grindstone.

Melissa and I had broken up in part because we no longer shared the same values. What I once saw as her happy-go-lucky and carefree spontaneity began to seem like irresponsibility. As I got more serious about my career, I wanted someone similar. That was part of what attracted me to Terri, the ambitious business major whose parents were not exactly thrilled by our relationship.

By my last semester when I graduated, I learned that I was on the dean's list: no more Cs. I could hardly believe it. After getting the good news, I went to a nearby playground with Terri. She was a diligent and organized student and I thought she was extremely intelligent. I also saw that she put a lot of effort into her own schoolwork.

As we sat on some swings, Terri told me, "You've got it. You can do whatever you want in terms of education." She looked me in the eyes to be sure I took it in. I knew she was going places herself. For her to say that about me really meant something. It was the first time I believed I could actually get my PhD. But before I could attend graduate school, I still had some deficiencies to remedy.

Soon I found myself spending twelve hours at a stretch in the lab, at least five times a week. Rob began to teach me to operate on the brains of the rats we were studying. After I got over my initial fear and disgust, I found that I was good at it. Soon I was basically doing brain surgery with ease, using a surgical suite that looked like it was set up for tiny dolls.

My undergraduate work also came at a time of tremendous excitement in neuroscience. That, too, inspired me, at times when my motivation began to flag. In 1990, as I mentioned earlier, Congress and President George H. W. Bush declared the 1990s to be the decade of the brain, calling for greater national focus on neuroscience to accompany the increased funding the field was receiving. It seemed like important new discoveries were being made every day. We thought we would soon find answers to the deepest and most difficult questions about thought, desire, and action, questions that had challenged the greatest human minds for centuries. I was studying the heart of the system that was said to provide pleasure and drive desire, a specific dopamine network in the center of the brain. We figured we were close to understanding how it worked.

I felt like I was really learning something, that this knowledge was important and vital. If we could understand dopamine, we would decipher desire and unlock addiction. The science itself was intoxicating. With enthusiastic encouragement from Rob, Don, and Jim, I was soon on my way to graduate school. The black kid who'd once been in the trailer for the learning disabled, whom his high school had relegated to business math and parking patrol class, was now on his way to a doctorate. I could now see a clear way out of the maze.

CHAPTER 11

Wyoming

Equal Rights

—WYOMING STATE MOTTO

It was a cold night in Wyoming, not the worst kind, where your face numbs in even a brief exposure, but still a stunning chill that a Floridian has no words to describe and no preparation to cope with. MH and my sister Brenda had braved the late-winter weather for a visit; I was then working on my graduate studies at the University of Wyoming in Laramie. Snow was everywhere. As the writer John Edgar Wideman observed in his book *Brothers and Keepers*, it snowed so much in Wyoming that it could make a grown man cry.

I'd earlier driven my mom and sister through the sleepy little town where I lived and then brought them to campus. I wanted to show them my lab. On most winter evenings, the campus was dark and desolate: most of the students and faculty quite sensibly did not linger outside. I started to select the right keys from a ring I carried and prepared to let them in. But MH was hesitant to go any further.

Despite the freezing temperature and our desire to get in out of the cold, I could see reluctance in her eyes. Her thickest winter coat offered little protection—but she was more frightened of entering the building than she was of the elements. She thought we'd get in trouble, possibly be arrested. Even though I had my own set of keys and had told her I worked there day and night, she remained concerned. Part of her still didn't believe that a black man could legitimately enter a university building at night and that her son was actually a graduate student who spent many long evenings doing scientific research in this alien place.

The moment stuck with me as a vivid demonstration of how my family and I had internalized racist tropes about "knowing our place." At this point, Brenda worked for Delta Air Lines as a reservations agent and her travel privileges were what allowed them to afford the visit. Like me, Brenda was starting to achieve some success in mainstream America, but every gain was hard-won and required ongoing struggle. We'd all had years of conditioning suggesting that a black person would not be accepted without suspicion in such a situation; the insidious nature of these unconscious cues shaping our feelings and behavior was crystallized in that moment for me.

My family had given me all the help that they could, but without the emotional and academic support of my mentors, girlfriends, and friends, I would never have been able to survive the transition to graduate school and ultimately get my doctorate. The social skills I'd learned in childhood had allowed me to get to this place; I'd need them more than ever to succeed here. No one—let alone someone from my background—could thrive here on their own.

As I'd advanced in my career, I had moved into environments that were progressively less black. Wyoming was the whitest. Both in terms of the wintry physical surroundings and the overwhelming sea of white faces on campus, it had the least color of anywhere I'd ever been. In fact, my time in the air force in England turned out to be the last time I worked in a genuinely integrated environment. As my scientific career moved forward, the number of black peers around me dwindled until frequently I was the only black person in the room. When I got my PhD in 1996, in fact, I was the only black man in America to receive a doctorate in neuroscience that year.

While Wyoming was blindingly white, however, its whiteness was different in character from that of UNC-Wilmington. There the campus had an overwhelming white majority in spite of being surrounded by a large black community and I experienced more overt hostility toward people who looked like me. In places like North Carolina and even New York, stereotypes about black people were often reinforced by what people saw around them: in Wilmington, for example, I'd often be the only black student doing research and involved in research-related functions, and most of the blacks on campus worked in low-level or service jobs, not academic or administrative positions. As I noted earlier, this is why many black Wilmingtonians referred to the university as UNC-White. Back east, white people saw blacks and maybe thought about rappers, poor people, or even criminals: their initial perceptions certainly weren't of students, let alone scientists.

But here in Wyoming, the large white majority simply reflected the actual population. And any blacks who were on campus were typically stars: they were athletes or outstanding students; they had no other reason to be in remote Wyoming. There were so few blacks that other people saw us almost as celebrities, and that seemed to allow them to consider us more as individuals and less through the lens of negative group stereotype.

Indeed, when I first visited the Laramie campus in early 1992, the man who would become my graduate mentor took me to a college basketball game. "That is probably the most black people you will see in one place, right down on that floor," Charles Ksir told me, indicating the players. We were surrounded by thousands of cheering white faces, some painted in the Cowboys' awful signature colors of yellow and brown. The crowd was enthusiastic. On a campus of around fifteen thousand people, there were probably a few dozen blacks, most of them members of the basketball or football teams.

Ksir, whom I would soon come to call Charlie, had been Rob Hakan's mentor during graduate school. Rob had encouraged me to apply to study with him at Wyoming and follow in his academic footsteps. As it turned out, it was the only graduate program in psychology and neuroscience to which I was admitted. While my grades were good and my lab work was stellar, my scores on the test typically used to determine graduate school admissions, the GRE, were abysmal—particularly on the verbal part. And I'd achieved the score I did only with lots of help.

Though it may not sound like it now because I've worked so hard on vocabulary, back in college I still didn't know as many words as were expected of someone seeking a PhD. My lack of exposure to mainstream language in early life was another obstacle I had to overcome. Rob had bought me word books and quizzed me on lists of new words about once a week. Jim had also helped expand my language skills. But I hadn't advanced enough by the time I took the GRE to have overcome the severe deficit with which I'd started, at least as far as could be measured on that standardized test. Unlike richer students faced with lower-than-desired test scores, I couldn't afford prep courses. I had to rely on my mentors and friends.

And Charlie immediately made me feel welcome in Wyoming. Soon he became one of the key nodes in the new social support network I built that enabled me to get my PhD. Charlie was a professor of psychology and was studying the effects of nicotine on dopamine at the time. When I visited, it was February, the deepest trough of winter. I walked past the booth set up to celebrate Black History Month—and noticed that its attendants were white people. I'd never seen that before; there was simply no black student available to do the job.

Charlie gave me a complete tour. As we walked through the campus bookstore, he pointed to a book that was prominently displayed, called *Black Robes, White Justice*. It was the autobiography of Judge Bruce McMarion Wright. He asked if I'd read it. I hadn't, but I did know that Judge Wright was better known in New York as "Turn 'Em Loose Bruce" for what the police and prosecutors saw as his lenient sentencing decisions. He was black and a prominent civil libertarian. Charlie used the book to start a conversation that let me know he had thought deeply about how race plays out in the United States and that his knowledge and intellectual interests extended beyond neuroscience.

This was important to me because I knew people would expect more from me than they would from a white person in the same position. For example, I would be expected to know something about why there were so few black neuroscientists or something about how to address the "drug problem" in black communities. The conversation with Charlie suggested that he knew this as well, and that was encouraging and reassuring.

During our walk and later back in his office, we talked frankly about race and justice in America. This was a topic that the white folks with whom I'd interacted back in North Carolina had always studiously avoided. And when it did come up, even my well-intended white mentors would often say things about how I should shape my attitude to be sure I was able to best take advantage of the opportunities I had. They never acknowledged how awful or disturbing it was that I continually had to confront the dilemma or that the fundamental problem was the racism, not my response to it. This made it feel like it was my own personal issue and it was an ongoing irritant.

In contrast, Charlie started by putting it all on the table. In essence he said, "It's there, I see it and I'm white, and it's not something wrong with you." He talked about his youth in Berkeley, California, during the days of the Black Muslims and how it was oh so easy to talk the correct liberal talk. But actually participating and working with others to try to do something about it: now, that was something else entirely. Charlie had engaged in repeated discussions with Black Muslims and had been called a "blue-eyed devil" for his efforts; he knew how to deal with racial and political conflict up close and personal.

I decided right then that, if accepted, I'd attend graduate school at Wyoming, and Charlie became my most important mentor there. I knew I could learn from him since he was so willing to be straightforward, rather than dodging unmentionable tensions or assumptions or dismissing the prevalence of racism itself. And so, when I did receive my acceptance letter in April 1992, I was eager to attend.

Indeed, in order to take Rob's advice about outworking those who might have other advantages, I decided to get an early start. Charlie hired me to work in his lab the summer before my first classes started. There I would perform the experiments I wanted to conduct for my master's thesis before beginning my course work in September. This research involved studying the effects of nicotine on dopamine in the nucleus accumbens, a region thought to be involved in the experience of pleasure and reward. This was a line of research that aligned with Charlie's own interests. I'd spend more time with rats, doing more brain surgery on them. I knew that I was well prepared to do the lab work.

I wasn't so sure about the classes, however. Thankfully,

*On my recruitment visit to the University
of Wyoming, Charlie took me skiing. It
was my first and last time skiing.*

before starting graduate school I spent a week in May with my
girlfriend Terri's father. He lived in Longmont, Colorado, and
taught me a critical lesson that paved the way for my graduate
school success. Terri's father had served in the military and was
an information technology consultant. He said that the most

important thing for me to do in grad school was to ask questions when there was something I didn't know.

I sort of nodded politely when he told me; it seemed obvious. Of course, if you don't know something, you need to ask about it. I'd always worked under that principle and had not been embarrassed previously to ask what might be seen as dumb questions. That had long been one of the keys to my educational success.

But he stopped me. He could see that I wasn't hearing him. "No, seriously," he said. "This is important. If you don't know, you have to ask." I suddenly realized why he was stressing it: he knew I might feel that since I was now a graduate student, I would have to start pretending that I knew things I really didn't know. I might be embarrassed, at this new level of recognized achievement, to admit ignorance. He was right.

And if I hadn't followed his advice, I probably never would have gotten my master's, let alone a PhD. With my background and the holes in my early education, there were many important things I didn't know. I had to be brave enough to ask what others might see as obvious questions. Not learning key things I needed to know for my work would be worse than possibly looking ignorant for a moment. And often, it turned out, other graduate students were similarly baffled by the "dumb" things that I thought I should have already known.

Indeed, this is why teachers often say there are no dumb questions: sometimes the most important discoveries come from questioning seemingly axiomatic assumptions. One such assumption during my graduate training was that dopamine was the "pleasure" neurotransmitter, and that drugs like cocaine and nicotine produced pleasure by increasing the activity of this neurotransmitter in the brain. Seminal evidence for this view came from studies of rats trained to press a lever to receive intravenous injections of cocaine or nicotine. For example, when rats are given an opportunity to self-administer cocaine, they do so robustly. But when given a drug that blocks dopamine several minutes before having an opportunity to self-administer cocaine, well-trained rats initially work harder to receive cocaine injections but eventually give up, presumably because the dopamine signal is being blocked. Researchers interpreted the rats' initial burst of responding as an attempt to compensate for the lack of pleasure due to dopamine blockade.

With nicotine, however, under identical conditions, rats do not display the burst in responding; instead they stop responding immediately. Despite the fact that the rats' behaviors were different depending on the drug—cocaine or nicotine—many researchers' interpretation remained the same. That is, in both cases it was interpreted that the animals were no longer able to get the pleasure experience they'd come to expect, because dopamine was being blocked. My question was, if so, then how could both responding more and responding less be interpreted the same?

I never received a satisfactory answer. At best, someone would say, "Good questions." Later I began to realize that the dopamine-pleasure connection was far more complicated than the way it was being described.

The more I studied drugs, in fact, the more I learned about these types of basic inconsistencies in our ideas about them. Back then, however, I was simply excited to be part of the scientific conversation and didn't dwell much on it. I found a study partner early on—doing so would be another key to my success—and settled in to do the work. My graduate work consisted of not only research and coursework but also teaching undergraduate courses. During my first year of graduate school, I served as Charlie's teaching assistant for his Drugs and Behavior course.

I taught the course on my own during my final three years of graduate school. By the time I had completed my graduate studies, I'd gained plenty of teaching experience.

Another academic mentor inspired me as well during graduate school. Jim Rose was the director of the neuroscience graduate program and the most thorough scientist I have ever met. Charlie introduced me to him during my initial visit to campus, taking me to his lab where he studied newts. I had never even seen one of these small brownish green aquatic salamanders before. But the wide range of experiments that Jim was conducting on their behavior and brains impressed me. From the molecular level to neural network level, all the way out to behavior, he was systematically exploring stress and sexual behavior in this animal.

Jim wasn't just your stereotypical cerebral scientist, either. A former high school wrestler and track star, he kept himself in such great physical shape that, at twenty-five years my senior, he could outrun me when we worked out together. His tolerance to the altitude may have had something to do with it; nonetheless, he frequently left me behind and huffing. Jim showed me that you could be manly and be a scientist—and he and his wife took care of me emotionally as well as physically. Every week, I'd have lunch with his wife, Jill, at Godfather's Pizza, where she was so well known to the staff that they kept a bottle of her personal salad dressing in the kitchen.

Jim helped me negotiate the politics of the university, as well as teaching my neuroanatomy, neuropsychology, and neuroscience of sleep classes. He taught me how to give a scientific talk. His critiques of my work were so rigorous that I knew that if I passed the "Jim test," I was ready to present my data to the world.

In Wyoming, of course, I also continued to spend hours upon hours in the lab. Charlie later said to me, "I never had a graduate student before who was as dedicated and put in as much time all

Charlie, MH, and me on the day I received my PhD.

on their own. Other students were interested and so on, but they just didn't put in the hours, and they weren't as single-minded as you were. You were just so focused on getting done what you needed to get done."

Indeed, I knew I was well on my way to becoming a real scientist when I found myself working Saturday afternoons at the lab during football season. It was located not far from the stadium where the Wyoming Cowboys played, and every time they scored a touchdown, a cannon would go off, loud enough to be heard in the lab. I was still a huge football fan, so making the choice not to go to a big game that was so close was a real sign of dedication for me. I was just hungry for knowledge and scientific experience.

Of course, I also felt extra pressure to compete well as a black person in such a white milieu. As Charlie put it: "I've tried to evaluate, well, was your race a benefit to you or a hindrance?

And obviously, in some ways it was a little of each, probably. It may have opened some doors in the sense of having people willing to give you the opportunities. But I [also] got the sense that there was a lot of begrudging of your going farther than they thought you would." It was as though people were pleased with themselves for giving me a chance, but astonished when I demolished the stereotypes they didn't believe they still held by becoming a true competitor.

This was clear from early on during my time in Wyoming. An experience I had at a cocktail party illustrates one way the issue played out. Probably during my second semester, I attended a party at the home of one of the neuroscience faculty members. This faculty member and I had a contentious relationship; he was disliked by many of the students because his teaching was obtuse and we struggled in his class. To make it worse, he belittled students and didn't show any respect for us. In short, we thought he was an asshole.

He had been raised on Long Island and my success seemed to make him especially uncomfortable. He'd make remarks like describing someone as "so rich he had the black maid and the black butler—no offense, Carl," in a way that made clear either that he was oblivious or was blatantly disingenuous about his intentions. I was pretty sure it was the latter, but it was hard to tell.

The neuroscience faculty and students got together for drinks or dinner regularly, either in the lab or at someone's house. It was pretty much the only type of socializing many of us did: graduate school takes up virtually all of your time. That week, it was his turn to host.

At one point during the party, he took me aside and said he wanted to show me something. We walked upstairs to his bedroom, where he pulled out a big-ass .44 Magnum, with a long barrel. It was obvious that what he was really doing was trying to demonstrate dominance and masculinity. So I played along.

I oohed and aahed as he described the technical features of the gun and some of his adventures shooting. I said, "Wow, that's a cool-ass gun."

Then I added, deadpan: "But when you come to my place, I gotta show you my Uzi."

His jaw dropped. His neck turned bright red. He had no idea how to respond. He couldn't tell that I was simply one-upping him: his ideas about blacks were such that he believed it perfectly plausible that I kept an Uzi in my grad school apartment. So I just said, "Yeah, man, remind me next time and I'll show you my Uzi," and went back to the party. He knew I'd trumped him. Because he wasn't sure whether I actually was crazy enough to have an Uzi, he backed off in his antagonistic interactions with me since I'd shown him that I couldn't easily be played.

But that was just a taste of what I faced as I worked to complete my master's degree in psychology in preparation for getting my PhD. And a racial incident on campus soon spurred me to my first experience with real activism.

The event that set things off wasn't especially egregious. The campus newspaper, the *Branding Iron*, had run a naive, literally sophomoric essay claiming that affirmative action isn't effective and that black students are given an unfair advantage, to the detriment of whites. Few would have objected to the mere publication of the piece: college is a place for people to explore ideas and make arguments and free speech means that some offensive and inappropriate material will invariably result. The real problem occurred because the paper, which usually ran counterpoint articles, did not do so in this case.

A group of athletes and a few other black and Latino students came to me for advice about how best to respond. By this point, I was pretty well known among them at the university, since I spent time at the multicultural center, I attended as many athletic events as I could in support of the teams, and most of the black athletes had taken my Drugs and Behavior course. We ultimately agreed that what we wanted was the opportunity to publish a reply—and I figured that this would be easy to get and that would be that.

But when I met the student editor of the paper, he flat out refused. Unexpectedly, the interaction became adversarial. He declared that it was his paper and no one could tell him what to print. At that point, I went to the university president and described the situation, asking him to reason with the newspaper editor. He met with us and then with the editor, who wouldn't back down. In an attempt to mediate, the president offered us three hundred dollars to pay for a full-page ad on the back of the paper, where the students could place any statement they wanted to make.

Although this solution did not provide an equivalent editorial reply, simply a convenient commercial one, I told him that we'd take the money. We ran an ad calling for a boycott of the paper and describing the entire series of events. In the ad, we also said we had the support of the university president and the psychology department, although we hadn't actually gotten permission from the president or the department to state this in the ad.

All of this got people's attention, particularly in sleepy Wyoming. At the same time, we discovered that the *Branding Iron*'s budget was supported by students' fees, including ours. Yet there were no students of color on the paper's staff. And when we said we were going to peacefully occupy the administration's offices, the story got even bigger. Now the local papers, the local television stations, even National Public Radio picked up on it. Soon I was meeting with the governor, who was a Democrat, and being asked by Democratic Party leaders if I could represent the state at some meeting related to student leadership.

Along the way, we also had the usual activist struggles over strategy and leadership, and when I began speaking out on race-related issues, my relationship with some of the white folks around me changed. This made me more suspicious and distrustful than usual. Jim Rose gave me one of the best pieces of advice I've ever gotten as a result, saying that I should face each person anew. Rather than defensively assuming that my views or actions had altered the relationship, I needed to be open first and let the other person's actual reaction—not my expectations or apprehensions—determine my response. This mindfulness of the present allowed me to deal with the situation in front of me as it was, not as I thought it might be, and that helped me immeasurably in academia.

Ultimately, although we did not get our counterpoint published as an op-ed in the student paper, the students who had protested did become more politically active on campus. Just after that, Wyoming had its first black student body president and the student senate experienced a wave of elections of minority students. Many of them later went on to jobs at the university—but sadly, most did not stick with their early activism. As is often the case, once many people become a part of the system they once criticized, they are rewarded for behaving in a manner similar to those around them.

Nonetheless, I had learned that I could organize people to take effective action. I was continuing to grow and learn as a scientist. While I wouldn't be conspicuously politically active again until much later in my career, the experience was galvanizing and formative. I was learning not only that I could succeed in academia but also that I might be able to change it.

The most important relationship I began in Wyoming, however, was with the woman who would become my wife and the mother of my two sons. Robin and I first crossed paths when I served as graduate adviser to the psychology honor society there in 1992. She was a psych undergraduate at the time. Her intelligence deeply impressed me. In fact, I suspected she was smarter than I was. At age twenty-six, she already had undergraduate degrees in international studies and French.

Robin was white. She was also one of the most beautiful women I'd ever seen. Her style was striking. She always wore stylish hats and scarves, not just functional winter gear. While many of the students on campus looked like they'd just come in from feeding livestock at the ranch, Robin looked more like a Manhattanite, even though she had actually been raised in Montana.

She has olive skin and green eyes, with lovely deep brown hair. We were friends before we became involved, but when we took the same class together in 1994, I knew I had to make a move. After she brought a plant to my office as a gift, I could see that she was interested in me, too. Soon we were inseparable.

Unfortunately, not long after we first got together, I had to leave Wyoming. In the summer of 1993, I'd won a highly competitive minority fellowship to work at the National Institutes of Health: only one minority graduate or medical student in the entire United States was accepted each year. I hadn't even considered applying, but Charlie had insisted and I eventually relented.

And to my great surprise, I had won the chance to spend the summer working in Irv Kopin's lab. Kopin was studying the neurobiology of stress, trying to understand the neurotransmitters and metabolites involved. However, what was even more impressive was that the lab I'd worked in was where Julius Axelrod had done much of the work that won the 1970 Nobel Prize in Physiology or Medicine. Axelrod had solved key problems in understanding how brain cells talk to each other, discovering mechanisms involved in neurotransmitter storage, release, and inactivation. It was thrilling to work in the lab where these critical discoveries had been made—and even more exhilarating to be asked to return the following summer, after completing my master's, to do my PhD work there. That, however, would mean leaving Robin behind in Wyoming.

When Robin and I had first gotten together, it seemed simple. We were both intensely attracted to each other, physically and intellectually. But we were also both at a point in our academic careers where we had little time to devote to a long-term relationship. I assumed it would be a casual thing, a nice diversion from our academic pursuits.

However, over time, things got more and more intense. We spent all of our free time together—limited as it was by our work—and constantly talked. I opened up to her in ways I hadn't done previously, and she, too, shared a great deal of herself. We were always talking about books and ideas: she was the first woman in my life who gave me books as presents. She gave me *Washington Post* reporter Nathan McCall's *Makes Me Wanna Holler* as a master's graduation gift. I read it while enduring the lengthy and tedious ceremony.

Soon I could see that she really was the kind of woman I sought as a life partner, and I think she felt similarly. In most ways, she seemed perfect. Except, of course, for being white. I wasn't sure how to deal with that, even as I hated that it mattered.

It was fine to have a fling with a white woman in Wyoming—but I couldn't imagine making a family with one, given all the baggage that interracial relationships carried in the wider world. Together we read Derrick Bell's *Faces at the Bottom of the Well*, particularly the allegorical short story "The Last Black Hero," which tells the tragic tale of a black militant who falls in love with a white woman and faces the paradoxes of trying to fight for racial equality while living in the inequitable world as it is.

Like the activist in the story, I was uncomfortable envisioning a future with a woman who wasn't black. I thought about what little black girls would think as they saw so many of the most successful black men marrying white women. I wanted to be one of those success stories—but I didn't want to disappoint the people who looked up to me. I certainly didn't want to reinforce the image that black women weren't good enough for high-achieving black men.

And so, as I prepared to leave for the NIH, Robin could tell that something was up and we needed to talk. She drove me to a spot up in the mountains, with a majestic view of the wide-open sky. Darkness had fallen and the stars were out. They seemed like they were everywhere in the late spring chill as we sat in the car on the mountainside. We started to talk.

I didn't want to hurt her, but I knew that if we got much closer, that was inevitable. So I explained as kindly as I could what I'd been thinking. I told her that I didn't know if I could face my community and be the man I wanted to be if I was with a white woman. I stressed that it had nothing to do with her and that our relationship itself was wonderful. I didn't want to have to make this decision. But to my surprise, she understood immediately. She didn't want to let me go, but she didn't want to stand in my way, either.

I hadn't intended on breaking up with her, just talking it all through, but that's where we seemed to end up. It was painful, but we decided to stay in touch and be friends. I hated to do it—and hated that race was so inescapable—but I couldn't see a way around it. I left for the NIH believing that our relationship was over.

CHAPTER 12

Still Just a Nigga

To be a Negro in this country and to be relatively conscious is to be in a rage almost all the time.

—JAMES BALDWIN

egro Cocaine 'Fiends' Are a New Southern Menace"

That was the title of the "journal article" I'd discovered when I began trying to track down a reference from a paper I'd read about cocaine. I was looking for early historical reports of cocaine withdrawal. The authors had cited the reference with a disclaimer. They wrote: "Reports of patients with similar symptoms had appeared in the early 1900s, but because these reports were deeply interwoven with elements of racist hysteria they were never taken seriously." But I still wasn't prepared for what I found when I read the entire article.

Of course, I knew that such blatant racism was common even in the medical literature in the Jim Crow era, and that I couldn't hold historical work to modern standards. This was just science. If the author had accurately described cocaine withdrawal, it could be a useful citation, I told myself.

It was March 1996 and I was in the science library at the University of Wyoming, finishing up my PhD. My dissertation dealt with how nicotine's behavioral effects were influenced by changes in parts of nerve cells called calcium channels. For the opening of my thesis, I was required to describe the rationale for the experiments I'd done. That involved comparing the effects of nicotine to those of cocaine, and I wanted to cite relevant work about the influence of cocaine on human behavior. And since my education had shown me that if I had a particular thought, someone else had probably already considered the idea in depth, I went back as far as the leads would take me.

The paper that cited the provocatively headlined article had used it to support a claim that cocaine-related deaths and other problems had been described early in the drug's history. I wanted to see for myself what arguments it made. Though immediately offended by the language of the header, I was also excited because I'd never seen this paper cited before. If I could track it down, I might be able to find a very early description of cocaine to add to my work, which might impress my professors.

My first surprise came when I read the full reference: the "journal" in which the article had been published did not seem to be some august peer-reviewed medical publication. It was, instead, listed oddly as "New York," perhaps having been cut off by mistake. I can't recall how, but I eventually ascertained that what was meant was actually the *New York Times*, and, even though I now knew it was just a newspaper story published on February 8, 1914,[1] I decided to get a copy of the whole article.

I walked across the snowy campus to Coe Library, the university's main reference library. Old newspapers were stored there on clunky microfilm, not kept in the more specialized science library where I did most of my literature searches. I looked up the citation in a big bound index with a thick, worn cover. Then I requested the relevant reels of microfilm and watched them scroll blurrily across the reader's screen until I found the right frames. That was what research was like in the days before the Internet.

The first thing I could read besides the headline was the sub-head: "Murder and Insanity Increasing Among Lower Class Blacks Because They Have Taken to 'Sniffing' Since Deprived of Whisky by Prohibition."

I was surprised at how shocked I was to see that. I knew intellectually that such blatantly racist writings existed and that it was once acceptable to print such things in respectable papers, but it had always seemed abstract and distant to me. It was very different to see the words in black-and-white on the pages of the *New York Times*, the publication that to this day is seen as the "paper of record." It was as different as reading about slavery in a history book is from holding in your hand an iron shackle once used to bind a real human being. Or as different as learning about the Holocaust in history books, versus actually going to Auschwitz and seeing firsthand the shoes of the children killed there.

But what shook me even further was how similar the article was to modern coverage of crack cocaine in the mid-1980s. The author, who was a medical doctor, wrote:

> Most of the negroes are poor, illiterate and shiftless. . . . Once the negro has formed the habit he is irreclaimable. The only method to keep him away from taking the drug is by imprisoning him. And this is merely palliative treatment, for he returns inevitably to the drug habit when released.[2]

This rhetoric was unsettlingly modern. For example, recall what Dr. Frank Gawin told *Newsweek* on June 16, 1986: "The best way to reduce demand would be to have God redesign the human brain to change the way cocaine reacts with certain neurons." The message is that crack users are irretrievable, except for divine intervention. Of course, in 1986 explicit reference to race in such a context was no longer acceptable; instead, crack-cocaine-related problems were described as being most prevalent "in the inner city" and "the ghettos." The terms *inner city* and *ghetto* are now code words referring to black people.

Dr. Edward H. Williams, author of the "Fiends" article, went on to claim:

> [Cocaine] produces several other conditions that make the "fiend" a peculiarly dangerous criminal. One of these conditions is a temporary immunity to shock—a resistance to the "knock down," effects of fatal wounds. Bullets fired into vital parts that would drop a sane man in his tracks, fail to check the "fiend."[3]

In other words, cocaine makes black men both murderous and, at least temporarily, impervious to bullets. By the way, the author was describing the effects of cocaine taken by snorting it. Attempting to further bolster his case, the writer then added anecdotes from southern sheriffs, who claimed to need higher-caliber bullets to take down these black "fiends." He also contended that cocaine improves the marksmanship of blacks, making us even more dangerous to the police and society.

I began to wonder how many of the "truths" that I now thought to be obvious about drugs were similarly shaped by racial bias. And I soon learned that it was sensational reporting like this that had largely led first to state and then national prohibition of the currently illegal drugs. I read histories like David Musto's 1973 classic, *The American Disease: Origins of Narcotic Control*, which helped me to further understand that drug laws banning drugs like cocaine, opioids, and marijuana were based less on pharmacology and more on racial vilification and discrimination.

For example, between 1898 and 1914 numerous articles appeared in the scientific literature and popular press exaggerating the association of heinous crimes and cocaine use by blacks: the *New York Times* piece was not an exception, but an example. As Musto has detailed, "experts" testified before Congress that "most of the attacks upon white women of the South are the direct result of a cocaine-crazed Negro brain."[4] As a result, it was not difficult to get passage of the Harrison Narcotics Tax Act of 1914, which effectively prohibited the drug.

Before learning this history, I'd always assumed that the legal status of a particular drug was determined primarily by its pharmacology. However, I found that there were actually no sound pharmacologically rational reasons behind why alcohol and tobacco were legal, and cocaine and marijuana were not. It was mainly about history and social reasons, about choosing the drug dangers that would be highlighted to spur public concern and those that would be ignored. It seemed as if sound pharmacology was almost never considered or minimized.

Bans on drugs were inevitably preceded by hysterical coverage filled with scare stories about drug use by despised minorities, often immigrants and the poor. As Musto details, in the case of cocaine, the fears were linked to southern blacks. With marijuana it was blacks and Mexicans who were the bogeymen, and with opium it was Chinese railroad workers. In all three cases, sensational press reports were coupled with salacious portrayals of males of these groups using the drugs to facilitate rape or seduction or both of white women.[5] Even national alcohol prohibition had been passed with an aim at controlling the behavior of what the mainstream saw as frightening minority groups. In that case, it was primarily beer-drinking Germans and other poor immigrants in the run-up to and during America's involvement in World War I.

My skepticism about the nature of the drug problem slowly increased during my academic training. For one, under Charlie Ksir's tutelage, I had begun teaching a course on Drugs and Behavior, starting out as his teaching assistant. In the class and in the textbook he wrote that we used (I became his coauthor on later editions), myths about drugs were constantly discussed and debunked.

For example, in one lecture, I remember him carefully presenting data showing that cocaine-exposed infants fared no worse than those who had been exposed to nicotine during their mother's pregnancy. Another time, I remember Charlie calling the Office of National Drug Control Policy (ONDCP, better known as the drug czar's office) to ask for the source of some information. An advertisement they'd released had claimed that some high number of cocaine-exposed infants was born every minute. But when Charlie pressed the ONDCP representative for the citation, it became clear that the number was obtained by extrapolating from other numbers. At best, this wasn't an ideal strategy, and at worst, it wildly overstated the real statistic.

At first I found these facts hard to believe because of everything I'd heard about the dangers of crack cocaine. But I soon realized that I had nothing other than what I could now see as media hype to support my position. Jim Rose had pounded into my head the need to back all of my statements with rigorous empirical data, and when I started applying my critical thinking skills to what I thought I knew about drugs, very little survived.

Much of what we learn as scientists involves critically interrogating the methodology used to conduct research and trying to root out as many sources of bias as we can. The media, however, does not apply these methods to its reporting and therefore frequently presents an overly simplistic and a distorted picture.

Did we really now understand cocaine in a more sophisticated scientific manner—or had we just changed the language that we use about it in a way that hid the racist stereotypes that were so obvious in 1914? Starting in graduate school, I slowly began to question everything I thought I knew about drugs, in light of these disturbing parallels and the clearly racially driven origins of the drug laws.

An experience I had myself at the NIH, where I'd started work on my PhD after completing my master's degree in Wyoming, also made me think more about this. Located in Bethesda, Maryland, the main branch of the agency looks like the medical

Collecting data for my PhD research at the
National Institutes of Health (NIH).

center on a major university campus. It's a self-contained world, with dozens of boxy high-rise, hospital-type buildings and labs. It even has its own bank, the NIH Credit Union, which is located in Building 36. That was about a hundred yards from the main clinical center where I worked, in Building 10.

Walking over to the bank, I was your classic absent-minded scientist: my mind focused on the samples I was working on and the data I needed to collect rather than on my surroundings. At the time, I had joked with friends that I was afraid of losing my social skills because I was spending so much time alone or with rats—but I was actually a bit afraid that it might be truth rather than just humor. I was entirely wrapped up in my work.

As I left the credit union after depositing my paycheck or getting some cash, two men approached me. They were looking at me so intently as I came out of the door that my first thought was that they were gay men trying to pick me up. I was dressed in a dark purple sweat suit that was fashionable among young black men at the time and had my big laminated NIH ID hanging prominently from a lanyard around my neck. I had a bank statement in my hand. I noticed the men's intense stares, but at this point, I was still thinking about my lab work.

When they approached me, however, they identified themselves as police; the NIH campus was so large that it actually had its own force. One said to me, "A crime just happened and we want to know if you can help us." I said, "Sure, absolutely, whatever I can do." I had no idea that I was the suspect. I identified myself as a doctoral student conducting research and offered them my bank statement.

Nonetheless, the two officers told me that there'd been a strong-arm robbery near the bank and that the perpetrator was wearing dark clothing. That is all I was told. I assumed that the suspect was black but didn't learn this from the police. Nor was I told the suspect's height, weight, or any other identifying characteristics. What was apparent was that the two officers, who seemed to be in charge, were brown-skinned: a black guy and a Filipino.

Of course, it would have been rather stupid for a bank robber to return to the scene of the crime for another transaction—let alone provide a bank statement full of identifying information—but that didn't matter. Being a young black man wearing dark clothing was enough for me to "fit the description." Nor did it matter that the officers themselves were minorities. In many instances like this, because institutional racism is so pervasive in some police organizations, the behavior of minority officers is more egregious than that of their white colleagues, in part because everyone in the organization knows what gets reinforced (rewarded) and what gets punished. The risks of mistreating me are far less than those of mistreating a white counterpart, who may be the son or relative of some "important person."

The police asked: would I consent to walking in front of one of the campus buildings so the victim could try to identify me? They wanted me to participate in an impromptu one-man lineup, something that is notoriously unreliable. I didn't see any choice other than to agree. I walked toward the police cars that I now saw across the parking lot and was told that the crime victim was watching from one of the windows. They had me turn one way, then another so that the person could get a better view. After about twenty minutes, they let me go, saying that the victim hadn't recognized me. The whole thing was excruciatingly embarrassing, being conducted in the center of campus where any of my friends or colleagues could potentially have seen it.

By the time they let me go, I was in part relieved and in part working to tamp down my anger, something I'd had to become extremely skilled at by this point. I went to see my NIH mentor, but he didn't understand why the incident had affected me so deeply. He tried to use the comparison of himself—an elderly white man—being stopped by police in a black area of Washington and asked why he was there.

That made it only worse because it didn't reflect the reality. Like many blacks, I'd come to expect this sort of denial and minimization from white people—many of whom seem to see acknowledging racial injustice as an admission of guilt by association or a signal that their privilege is undeserved. But I still

*The NIH ID that I was wearing when
I was stopped by the NIH police
and subjected to an impromptu
one-man lineup.*

felt a bit betrayed by his inability to recognize my perspective, and worse than before I'd gone to see him.

There I was with my NIH ID around my neck and my bank statement in my hand and I was still seen as a likely bank robber who'd strong-armed a customer. Or a "Negro cocaine fiend," for that matter. Here in the United States, I was still just another nigga, no matter how many hours I had put into studying or conducting my experiments. When I met with Levon Parker, a black man who was director of student programs at the agency, and Leroy Penix, a black neurologist whom I sometimes shadowed on his rounds, they were upset but not surprised. The black professionals whom I respected didn't talk about it publicly, but they'd all had the same kinds of experiences. It hit me hard why some of the blacks I knew at the agency called the place the "plantation." The overwhelming majority of the scientists were white and most of the support staff was black.

Parker contacted Harold Varmus, who was then the head of NIH. I was asked to meet with the director to discuss the situation. Soon my phone was ringing off the hook with people trying to pacify me and prevent what happened from being publicized and becoming an embarrassment to the agency. They wanted me to meet with the NIH police and tell them what they should be doing better, even though I had no qualifications for this task other than being black. Even then, I recognized it as a token response.

Since I was just starting my PhD, I didn't want to attract this kind of attention to myself, either. I spoke with Varmus on the phone (he was traveling) and met with his staff. I told them what I thought, but I realized that without public exposure and specific policy alterations, these incidents rarely lead to change. It was like the "beer summit" that President Obama later had with the Cambridge, Massachusetts, police officer who arrested

Harvard professor Henry Louis Gates Jr. after he was seen trying to get into his own house. Instead of tackling and revising
the policies that produced these institutionally racist results, the
events were symbolically addressed as isolated misunderstandings. The system that produced them was left untouched.

Although I'd tried to cut my ties with her, the "breakup" that
Robin and I had negotiated didn't last. Less than a month
went by before I realized how much I missed her. I began to think
we'd made a big mistake. I had few friends in Washington, D.C.,
and none was as close as her. Although she was pursuing her PhD
in clinical psychology in Wyoming, we spoke on the phone frequently and her support after my near arrest was rock solid. She
helped me write the letters to the NIH officials that I sent as events
unfolded. While I was dating other women, I began to yearn to
see her again. I invited her to visit and she accepted.

I'll never forget the dress she wore when she came to Washington on June 10, 1994. It was a bold, brilliant blue and had
a demure white collar. Our reunion was passionate, intense.
Although I wouldn't know it for a few months, we conceived
our son Damon that night.

Still, when she called several weeks later to tell me that she
was pregnant, I didn't know what to do. I remained ambivalent
about creating a family with a white woman and deeply concerned about the issues Derrick Bell had so aptly described as
making these relationships so fraught. But one thing I knew I
didn't want to do was leave a child fatherless. As the pregnancy
progressed, I knew I had to make a decision about whether I'd
return to Wyoming and be with Robin.

And so, when Damon was born on March 13, 1995, I was
right there in the delivery room. I watched in awe as Robin per-

severed through hours of labor. We had a large private room at Iverson Hospital in Laramie, Wyoming. She had wanted and achieved a drug-free birth.

I'd brought my CDs to play soothing music for her and we listened to Bob Marley as the contractions got closer together and became more intense. I was overwhelmed by Robin's beauty and grace throughout the whole messy and sometimes frightening process. Indeed, moments before Damon was actually born, I'd seen a look of concern flash in the doctor's eye as he discovered that the umbilical cord was wrapped around our baby's throat—but he didn't tell us that this was what had occurred until the boy was safe in our arms. I couldn't believe I was a father. It was beyond anything I'd ever experienced.

I'd never before felt as happy or close to anyone as I did to my little family when we first held Damon. The responsibility we had for this tiny, brand-new life felt like both a blessing and an almost unbearable burden. I had been reading John Edgar Wideman's *Fatheralong*, which emphasized the difficult task faced by black fathers in protecting their sons. I was humbled by the challenge I faced, keeping a black boy safe while he grew up in the America I knew.

I also couldn't believe that they were letting people as inexperienced as we were take this fragile creature home with us. At the same time, I wanted to give him everything I'd always wanted from my father. I realized I had no clue what I was doing. I knew my life would have to change.

For one, I recognized that I had to get serious about our relationship and resolve my internal interracial-couple conflict. I wasn't yet sure exactly how to do this, but I knew for certain that I wanted to raise my son right. I wanted the security of a two-parent home for my baby. I certainly didn't want any child of mine to have the kind of chaotic home life that I'd experienced.

Robin and Damon in Wyoming while I was in D.C. studying at the NIH.

I ultimately decided not to stay at NIH, where I'd planned to complete my PhD. Instead I'd return to Wyoming to do it, so that I could be with Robin and our son. We would ultimately get married there, three years after Damon was born, on May 23, 1998, in a simple ceremony at Wyoming's Newman Center, following Robin's Catholic traditions. But first, I had to go back to Washington shortly after Damon was born to wrap up my research before I'd be able to return to Wyoming to get my degree.

While at a D.C. Metro stop waiting for a train, I began what turned into a lengthy conversation with a machine technician who was working in the station, repairing the ticket vending equipment. I had complimented him on his dreadlocks, thinking that he wore them as part of the Rastafarian religion. For years I had considered growing dreads myself, but I'd always held back because I believed that it was disrespectful if you weren't a part of that religion. I also did not want to be seen as faddish or simply following the crowd: that was not how I wanted to live.

But this man said that for him, wearing dreads was a way of showing homage and respect, even though he wasn't religious. That resonated with me, as did his self-assurance and thoughtfulness. By the time I left, we were no longer strangers. And I decided right then to grow my hair. It would remind me that I could be myself and be a conscious spirit, no matter what other people might decide a scientist should look like. It would connect me both to my heritage and my new son. It felt right.

I found myself thinking about this and about Damon's future a few months later, when Louis Farrakhan gave the keynote speech at the Million Man March on October 16, 1995. I'd been unable to attend since I was back at work on my research in Wyoming by that time, but I watched on TV as I minded Damon. Here were hundreds of thousands, perhaps more than a million black men. They were leaders, businessmen, professional people like Barack Obama (who attended himself), mainly middle class, and virtually all employed. It was inspiring to see.

And yet, the rhetoric was tightly focused on hard work and responsibility, on pulling ourselves up by our bootstraps and supporting our families. No demands were being made of Congress; no delegations sent just a few streets over to meet with our senators and representatives. Here were people who had done what we were supposed to do—not people who were uneducated or unmotivated—and they still didn't get it. They had bought into the mainstream narrative that we ourselves were the prob-

lem; that we were to blame for things like the selective enforcement of drug laws, the underfunded schools, and biased hiring
that hindered so many.

These were men who still tried to fit into a country that didn't
want to recognize their contributions. They were all folks who
could still be put in the equivalent of a lineup outside a bank,
while carrying a statement and a photo ID identifying them as
a scientist at the premier government health research institution
in the world.

It infuriated me, but that's what I realized my son would
soon face. A world where even in the most clear-cut situation,
someone with our skin tone could still be seen as a "crackhead"
just because he dressed a certain way—or to use the language of
an earlier wave of drug hysteria, a "Negro cocaine fiend." And
all of this made me think a lot more critically about my research
and about how to think about drugs.

CHAPTER 13

The Behavior of Human Subjects

It is not heroin or cocaine that makes one an addict. It is the need to escape from a harsh reality.

—SHIRLEY CHISHOLM

Robert sat on a hospital bed, surrounded by about a half-dozen people. He was a tall, slender, light-skinned brother with a goatee and short hair, in his early thirties. He was reclining in a typical, austere single room, with a small window and the usual pale and sterile hospital decor. At the center of the group was Dr. Ellie McCance-Katz, the woman who had recruited me to a postdoctoral position in Yale University's Psychiatry Department.

A short, fortyish woman with auburn hair, Ellie led the team. A nurse and another doctor monitored Robert's blood pressure and other vital signs. A female research assistant and I were also clustered around Robert as he slowly received an intravenous injection of cocaine. It was December 1997.

Postdoctoral work is an important step in scientific training, which, if things go well, can lead to the ultimate academic prize: a tenure-track job at a reputable university. My Yale postdoc was also my first experience of studying the effects of psychoactive drugs on a human being. It was exciting to finally get to do this work.

Over time, I'd come to see the limitations of the animal research that had been my initiation into neuroscience. For example, there's a phenomenon seen in animals, called sensitization, that occurs when they are given stimulant drugs like cocaine. Typically, when rats take a drug repeatedly, they become tolerant to its effects and a higher dose is needed to reproduce the initial response. But with some effects of stimulants, animals actually become more sensitive to the drug and they have a bigger response to a smaller dose than they did at first: the opposite of tolerance.

In humans, this sensitization was said to cause addicted stimulant users to become more paranoid and anxious over time. However, that result isn't seen consistently in human drug users and it isn't seen when stimulants are used therapeutically, which suggests that it is not an important pharmacological effect for people. As I continued to study drugs, I found many similar phenomena that just didn't carry over. It all made me think that in order to discover what I really wanted to know about drug use, I'd have to study it carefully in humans.

Robert was an affable, handsome man. Dressed neatly but casually, he didn't look overly thin or sickly: there was nothing to suggest to anyone who saw him that he was a crack cocaine user. While we were blinded as to the dose of drug he was receiving and to whether it was a placebo, cocaine, or a cocaine-related compound called cocaethylene, I soon learned to tell when he got a decent dose of drug. Then all he wanted to do was talk. He'd go on and on, sometimes describing how cocaine gave him insight and creativity.

Our study was designed to compare the effects of IV cocaine
to IV cocaethylene, a compound that is created in the body when
cocaine and alcohol are taken together. At the time, there were
concerns that cocaethylene was more potent and more dangerous
to the heart and blood vessels than cocaine taken alone. Under
carefully controlled conditions, we wanted to learn whether this
was true when the drug was given to healthy people who typically used cocaine and alcohol together.

I recognize that some may question the ethics of giving drugs
like cocaine and cocaethylene for research purposes. Over the
course of my career, however, I have come to the conclusion
that it would be unethical *not* to conduct this type of research,
because it has provided a wealth of information about the real
effects of drugs and the findings have important implications
for public policy and the treatment of drug addiction. From this
study, for example, we found out that fears about the dangers
of cocaethylene were not supported by evidence. Cocaethylene
turns out to be less potent than cocaine.[1] It actually has less of
an effect in terms of raising heart rate and blood pressure than
does cocaine itself, meaning it probably carries less risk for heart
attack or stroke.

Back in 1997, when I started working on this study, I still
had many misconceptions about drugs myself. Like the idea that
cocaethylene was a major new threat, my other hypotheses were
being repeatedly contradicted by the data during my graduate
and postdoc studies. I'd had a previous postdoctoral appoint-
ment at the University of California, San Francisco, in 1996,
which I'd received right after graduating from Wyoming. I had
been eager to start studying human drug users and I knew I'd
have a chance to do so at UCSF.

But in California, I wasn't able to study people actually taking drugs in the lab: the researchers I worked with were focused

on drug craving, which was supposed to drive addiction. These scientists didn't study the effects of drugs themselves; they examined only what drug users were reporting about their desire for them. I rapidly discovered that craving wasn't as important as I had initially thought. This was another step in the evolution of my thinking about drugs.

The problems with craving first became clear when I interacted with real people who had sought help for addiction. To try to understand their desire for drugs, I had become a facilitator for group sessions required of the patients in a methadone program. Almost immediately, however, I began recognizing that I had much more in common with them than I'd expected. Although they did discuss drug-related issues, unless they were prompted, craving wasn't their primary concern. The patients' real issues were mainly related to practical things like the high cost of housing and other essentials. That was something I'd had a very acute personal experience of as I started my postdoc.

It had been so hard for me to find an affordable place to stay in the Bay Area that I'd actually spent the first several weeks of my postdoc sleeping in my office. This was one of the many frustrations I experienced during my postdoctoral training that sometimes made me seriously question my desire for a future in science. Postdoctoral work is critical to a scientist's career, but even now in 2013 it pays only $40,000–$50,000 a year. Back then the salary was a meager $19,000–$24,000. I understood what these men and women in treatment were going through, trying to survive on not much money and manage their work and relationships. I'd thought these drug users were going to be much more different from me than they actually were.

Instead, I found that people with addictions weren't driven only by drugs. Moreover, they weren't any more antisocial or criminal than people I'd grown up with, many of whom rarely or never got high; in fact, their behavior wasn't much different from what I'd engaged in myself with my friends back home. They didn't seem overwhelmed by craving: they basically sought drug rewards in the same way that they sought sex or food. I began to see that their drug-related behavior wasn't really that special and to think that perhaps their drive to take drugs obeyed the same rules that applied to these other human desires. The notion that addiction was some kind of "character defect" or extreme condition that created completely unpredictable and irrational actions began to seem misguided.

And when I heard lectures by addiction researchers who studied animals, I began to realize how they extrapolated from extreme situations in ways that created a caricature of addiction. One researcher talked about how you could leave a hundred-dollar bill in the room and "you or I wouldn't take it" but a drug addict always would. They talked about humans in simplistic ways that, ironically, lacked the careful qualifications they always included in their discussions of animal research.

Later, I also came to see how our distorted images of addiction played out in the attitudes the researchers had toward the study participants at Yale. For instance, Robert's musings on how cocaine made him more focused and creative were discounted as drug-induced drivel—and yet studies of the impact of cocaine on concentration do show that it can improve alertness and concentration, exactly as he claimed.

Other experiences led me to see even more similarities. David, a thirty-five-year-old Italian-American construction worker, also participated in the cocaethylene research. He once described his experience to me of the day he was recruited to participate in the study. He'd seen an ad in a local alternative newspaper, seeking frequent cocaine users willing to be in an experiment in which they might be administered cocaine. They had to be otherwise healthy and willing to live in the hospital for two weeks. If accepted and if they stayed the whole time, they'd be paid a thousand dollars upon completion.

We'd interviewed David and determined that he was an appropriate candidate. Then we arranged for him to get a physical at a clinic at Yale–New Haven Hospital. The building had a strange address—it was 950½ or something like that, which sounds distinctly fishy. As he left our facility and sought this bizarre address, he noticed that there were several police cars parked outside.

That made him anxious. But he did want to participate in the study and possibly make some money, so he persevered. When he got near to where he thought the address was, however, he saw police outside that building, too. He began thinking that we'd set him up, that when he got inside and asked about the study, he'd be arrested. He walked around the facility a few times, trying to figure out what to do and whether he should even ask someone about the weird address. Maybe asking for that number would be the cue for the police to arrest him?

From the perspective of a nonuser of illegal drugs, of course, this sounds like sheer paranoia. When I told the story to other people working on the study, they laughed knowingly about how cocaine can make users paranoid. But from David's perspective, there was nothing irrational about his fears. He was involved in illegal activity. Police actually were engaged in an intense war on drugs. Tens of thousands cocaine users had been arrested. And we had all seen those movies or TV shows where lawbreakers are lured to some building by promises of a prize of some sort, only to be arrested for some earlier crime.

David had been asked to go into a government building and admit his drug use, which is a crime, in order to supposedly get paid to possibly take an illegal drug. His worries were an understandable response to his experience in the cultural setting in which it took place. While cocaine and marijuana can certainly increase these kinds of fears, anyone engaging in illegal activity does need to be cautious if he wants to avoid getting caught.

It became increasingly clear to me how our prejudices about drug use and our punitive policies toward users themselves made people who take drugs seem less human and less rational. Drug users' behavior was always first ascribed to drugs rather than considered in light of other, equally prominent factors in the social world, like drug laws.

And in reality, virtually all of us sometimes find ourselves in situations where we persist in behavior despite negative consequences, just like addicted people do. Most people can't stick to a diet, many continue to eat fatty and sweet foods when they are gaining weight, or have had periods of heavy drinking or stayed in bad relationships and ignored the negative results, which is the same pattern of behavior seen in drug addiction. Sure, there are extreme cases where addicted people commit absurd crimes— but there are plenty of equally stupid crimes planned or committed by people who are stone-cold sober.

I thought about my friends and family back home and where they'd wound up while I was working my way up in academia. I considered behaviors that were impulsive and often seen as associated with alcohol and other drugs. I myself had shoplifted and stolen batteries and sold drugs. But while I had plenty of less than perfect qualities, I had no addictions. Many of my siblings and cousins also engaged in petty theft as teenagers, but again, this was usually unrelated to their alcohol or other drug use or lack thereof.

In my immediate family, three of my five sisters had had teenage pregnancies. One of my sisters did become a heavy drinker (although she nonetheless always met her occupational

and family obligations). And she had her first child at age nine-
teen but married the father a few months after the child's birth.
They are still together. But she is not the sister who stabbed a
woman in a fight over a man and was later stabbed herself in a
similar situation. The sister who got into those altercations does
not have substance abuse problems.

One of my sisters' husbands was arrested in connection with
a deadly shooting but not convicted—but that is not the brother-
in-law who went to rehab for crack cocaine abuse. And the in-
law who did have a crack problem? He went on to get a job in
plumbing, has a house twice the size of mine, and is a loving
father and husband.

Where was the connection between drugs and problems
here? Among my family—just as I was beginning to understand from the research as well—the link between addiction and
other forms of dysfunctional behavior was not as prominent as
the stereotypes suggest. In some cases, alcohol use or its after-
effects exacerbated violence: for example, when my father beat
my mother. Some of my cousins had struggled with crack. But
illicit drugs and addiction were far from the greatest threats to
our safety and chances of success. There seemed to be at least as
many—if not more—cases in which illicit drugs played little or
no role than there were situations in which their pharmacological effects seemed to matter. And if the drug highs themselves
didn't explain behavior, for me that meant behavior related to
lack of drugs—that is, craving—was even further away from
allowing us to predict it.

I had left my San Francisco postdoctoral position disillu-
sioned by the whole concept of craving. Some addicts certainly
reported drug craving: there was no doubt about that. But it didn't
really predict whether they relapsed, according to the majority
of research. Sometimes people would report severe craving but

not use drugs; other times, they'd use drugs in situations where they said they'd experienced no craving at all. It seemed to me that it would be much more useful to study people's actual decisions about whether to take drugs, rather than focus so much on what they said about what they wanted or craved in some hypothetical future. That's why I responded with enthusiasm when Dr. McCance-Katz had suggested I do a postdoc with her at Yale.

Although I didn't get to study drug-taking decisions at Yale, at least with Dr. McCance-Katz, I was able to observe people's behavior while under the influence, not just their ratings of their desires to use drugs. That brought me closer to the types of experiments I really wanted to do so I could understand the real effects of drugs, not just our projections of them.

In order to find people to participate in our research in New Haven, I also had to interview many drug users. At the time, I wasn't even making a distinction between drug use and addiction. Despite what I was starting to learn, I still thought all illegal drug use was problematic and that most people engaged in it were headed for addiction if indeed not already there. I didn't distinguish between addictive use that interferes with major life functions like relationships and work, and controlled use that is pleasurable and not destructive.

Like the addicted people I was studying, I was influenced by my social milieu. Everyone around me in the addiction field acted as though pathological use was more common than controlled use. Certainly if you read the scientific literature unskeptically, this is the impression with which you are left. Consequently, when I interviewed users at this time whose lives seemed unscathed by their drug use, I figured I just hadn't yet become skillful enough to ferret out their denial. After speaking with dozens of them, though, I started to think twice. Maybe I wasn't the one who was wrong.

I thought back on what I'd learned about behavior and how it is affected by punishment and reward, going back to B. F. Skinner. Were drugs really that different from other reinforcers or pleasures? I looked at the existing data on that question. In the animal research, the graphs representing how hard an animal is willing to work for a food or drug reward were almost identical: make access easy and provide few alternatives and animals will definitely eat a lot of sweet or fatty food or take a lot of cocaine or heroin.

However, the harder they have to work for any reward—whether it's a natural pleasure like food or sex or a more artificial one like drugs—the less of it they will tend to seek. This is true whether the animal being studied is a mouse, a rat, a monkey, or a human being. And both in humans and in other animals, these responses will vary depending on the presence of competing reinforcers.

For example, studies have found that when rhesus monkeys have to repeatedly press levers to get either a cocaine injection or a highly desirable food (banana pellets), their responses vary with both effort and dose. Quite sensibly, the monkeys will work harder to get a higher dose of cocaine and put in less effort for a lower dose or placebo. They will also choose larger quantities of banana pellets over smaller doses of cocaine. Even at the highest dose of cocaine offered, these animals will never choose cocaine over banana pellets more than 50 percent of the time.[2] Addictive behavior follows rules and is shaped by situations just like other types of behavior. It's not as weird or special as we make it out to be.

You may say, "Yes, that's fine with a drug like cocaine that doesn't produce obvious withdrawal symptoms. But what about a drug like heroin?" Indeed, physical withdrawal symptoms can be seen in chronic opiate (for example, heroin, morphine) users if they abruptly stop drug use. The symptoms usually begin about twelve to sixteen hours after the last heroin dose and look something like a case of the twenty-four-hour, or intestinal, flu. Most of us have experienced these symptoms at some point in our lives: nausea, vomiting, diarrhea, aches, pains, and a general sense of misery. While this condition is most unpleasant, rarely is it life-threatening or accurately depicted in films that suggest the sufferer is on the verge of death.

Throughout the 1960s, drug addiction was defined solely on the basis of the presence of physical dependence (a withdrawal syndrome). About that same time, a group of researchers began publishing findings that questioned this dominant view. They reported that: (1) monkeys would begin and maintain lever-pressing for opiates without first being made physically dependent; and (2) monkeys who had given themselves small amounts of a drug and who had never experienced withdrawal symptoms could be trained to work very hard for their opiate injections.[3] More recently, researchers have demonstrated that monkeys' lever-pressings for heroin injections do not correspond with the timing or severity of their withdrawal symptoms.[4] These findings, along with others, underscore the notion that physical dependence isn't the primary reason for continued drug use.

I started to put these ideas together as I was trying to make my way in academia and dealing with a very unpredictable experience of reinforcers and punishers of my own. Although research careers are rarely presented this way when we are trying to attract youth to science, the reality is that the field is intensely competitive and many highly qualified people do not wind up with tenure-track jobs or even jobs in industry that take advantage of their skills. At UCSF and then even more so at Yale, I came face-to-face with the fierceness of this competition. It was extremely demoralizing at times.

This fight for status was worse than what I'd seen on the street or on the basketball court, where it was at least clear when people were competing and what territory was in dispute. In academia, no one said anything to your face: it was all sneaky stuff, all easily denied or explained away as a "misunderstanding" or "miscommunication." Men didn't fight like men; they stabbed you in the back instead. The rules were actually clearer and easier to follow in the hood. But one of the true advantages of my background was that it made me sensitive to social signals, no matter where I encountered them. I was able to learn those used in academia and use them to win, even on such a convoluted playing field.

Nonetheless, there were definitely times when I came close to giving up, when the low salary and grueling work hours with no guarantee of a definite payoff wore me down. The work at UCSF had been disillusioning: as James Baldwin had put it, when you learn a craft well, you get to see its ugly side, and that's what happened to me, starting there. I felt that the research we were doing on craving was poorly conducted and not productive, that the link between what we were measuring and what happened in real-world drug-using settings was not strong enough to matter. Dr. McCance-Katz was at UCSF on sabbatical at the time and I mentioned these concerns to her, which is how I got invited to do my second postdoc, at Yale. Even there, however, I still had no clear path to that elusive goal of a real job, a permanent tenure-track position. I wasn't sure I'd ever be able to support my family doing the work I loved. And now, I sometimes hated it. A job at Walmart started to look good by contrast.

To make matters worse, after only months, I learned that Dr. McCance-Katz was soon going to be leaving Yale, which meant my job there would end as well. The viciousness and underhandedness of the competition I experienced during this postdoc was beyond anything I'd ever been faced with before. For example, when I learned that Dr. McCance-Katz was leaving Yale to accept a job elsewhere, I met with a senior member in the department who promised me a faculty position within the department. Later, when I attempted to follow up on the position, this person claimed to have no recollection of our previous conversation, saying that I must have misremembered.

Fortunately, it was at this point that I met Herb Kleber, who was then the director of the division on substance abuse in the department of psychiatry at Columbia. I had a friend who worked with him and said that his program at Columbia was going to be expanding. She introduced us at a scientific meeting and he tried to recruit me with the promise of a faculty position. I was especially excited about the idea of working at Columbia because his wife, Marian Fischman, studied crack cocaine administration in humans. She'd published a paper in the prestigious *Journal of the American Medical Association* showing that crack and powder cocaine were pharmacologically indistinguishable.[5] I eagerly prepared to visit New York for my interview.

However, when I met with Marian, virtually the first thing she said was "I don't know what Herb told you, but we don't have a faculty position. We can only offer you another postdoc." Given the amnesia I was starting to see at Yale, I ultimately agreed to do a third postdoc at Columbia. I didn't know when this job limbo would end or for how long I could stand it. I certainly wasn't receiving the rewards of a scientific career that had been expected.

Marian, however, promised that she would do everything she could to help me get a permanent position. She was true to her word. It was at Columbia that I would ultimately get a tenure-track job and reach tenure itself. And in my research there I began finding, as I'd suspected, that humans do respond to cocaine quite similarly to how they respond to other reinforcing experiences. Like the rest of us, people who are addicted to crack cocaine are sensitive not only to one type of pleasure but also to many. While severe addiction may narrow people's focus and reduce their ability to take pleasure in nondrug experiences, it does not turn them into people who cannot react to a variety of incentives. I began the work that illustrated this as a Columbia postdoc, a job I held from September 1998 through June 1999.

In the study I briefly described in the preface to this book, cocaine users were given a choice between various doses of cocaine and various amounts of vouchers for cash or merchan-

Marian Fischman's research group when I arrived at Columbia in 1998. From left, Marian is the fifth person standing. Herb Kleber is seated next to me.

dise.[6] On average, on the street, our participants spent $280 a week on cocaine. These were not casual or irregular users.

Our procedure worked like this. First, we recruited frequent crack users through ads in the *Village Voice* and from referrals by other users provided by those who replied to the ads. Then we screened the volunteers for health problems that would ethically preclude their participation in cocaine research (for example, heart disease). We also screened their urine to ensure that it was positive for cocaine, though we did not reveal that we were confirming their use in this fashion.

Those who were cleared to participate were paid to stay for two to three weeks in a ward at Columbia-Presbyterian Hospital in Harlem (now New York–Presbyterian). Before we did any of this, of course, we'd applied for and received special licenses to work with illegal drugs on human subjects and been cleared by an ethics committee called an institutional review board (IRB). Then we obtained the cocaine from a pharmaceutical company, keeping it locked in the pharmacy with other controlled substances, using careful procedures to account for all of it.

On days participants were scheduled to smoke cocaine, each one would sit in a small room with a computer at a desk, where we could observe them through a one-way mirror. A nurse was in a nearby room, monitoring her or his vital signs and lighting the crack pipe when cocaine was chosen. When they smoked crack, participants were blindfolded so that they couldn't see the size of the rock they were getting. We didn't want them to have visual cues that might amplify or diminish their expectations about the hit.

At the very start of each day, before having to make any choices, participants had a "sample" trial. That meant that they were allowed to try the dose of cocaine we were making available that day and to see and hold the cash or merchandise vouchers on offer. Both the researchers and the participants were blinded as to whether actual cocaine or placebo was placed in the crack pipe. After the user had sampled the day's dose, he or she would participate in five "choice trials," spaced fifteen minutes apart. When a choice was available, an image of two squares would appear on the computer and the participant had to either click the left (crack) or right (voucher) side of the mouse to indicate their choice.

In order to actually get the drug or voucher, they then had to press the space bar on the keyboard two hundred times. During their first four choice sessions, the choice was between a voucher for five dollars in cash or the day's cocaine dose; during the last four, they had the choice of the dose or the five-dollar merchandise voucher.

Again, the results were similar to those seen comparing different rewards in the animal literature and in earlier human trials. When larger cocaine doses were available, users almost always chose cocaine over the cash or merchandise voucher. So far, this was congruent with the idea that addiction makes people place drugs first. But the rest of the data demolished that theory, showing that lower doses were often resisted. Despite the popular conception that addicted people will choose any dose of drug over any other experience—especially once they've already had a taste of it to kindle their craving—this is not what we find in the lab. Even around drugs, addicted people are not simply slaves to craving. They can make rational choices.

This was the case even though the alternative in each choice had only a maximum value of five dollars. In total, our participants could earn up to fifty dollars each day by participating in two complete sessions, which was a significant sum given their usual low income. But if the "first hit produces irresistible craving" theory were true, any dose should have had infinite value during the moment of choice. The cocaine users shouldn't have been able to think beyond the five dollars to the fifty—or about the particular dose, if the idea that people with addiction are totally out of control once they start using drugs is true.

Nonetheless, on average, users in our studies smoked two fewer doses of cocaine when the alternative was cash as opposed to merchandise.[7] This meant that cash money was 10 percent more effective than vouchers in suppressing cocaine use. The conventional wisdom about addictive behavior being completely irrational couldn't at all account for this result. If people addicted to cocaine always went for drugs no matter what, there should have been no difference.

Because our findings were so different from what most people have been taught about drugs, critics sometimes argued that they only really showed that these crack users were saving their money to buy more cocaine on the street later. That itself, however, doesn't even support the conventional view of addiction. Weren't addicted people supposed to be unable to resist drugs that were in front of them and be incapable of saving up for drugs or anything else later? And why would someone turn down pure pharmaceutical cocaine in a legal setting in favor of possibly being beat on the street and getting stepped-on (adulterated) drugs illegally in the future? *That* would truly be irrational under the logic of the idea of addiction as something that "hijacked" the brain and took control of the will in favor of immediate drug-seeking.

Alternatively, some folks predictably claimed that the users we recruited "weren't really addicted." People who were genuinely addicted would never have turned down free crack cocaine, they said. If we'd studied participants with genuine drug problems, they argued, we would have had very different results. Our participants, however, clearly had arranged their lives around crack. They weren't rich folks who had an extra few hundred bucks a month to spend on cocaine: they typically had unstable living arrangements and few or nonexistent family ties. Many had been convicted of crack-cocaine-related crimes and all had tested positive for cocaine on multiple occasions during the screening process. Most could tell you where to get the best and most inexpensive cocaine in the city. If this wasn't "real" addiction, what was?

The more I studied actual drug use in human beings, the more I became convinced that it was a behavior that was amenable to change like any other. So why did it seem so intractable in neighborhoods like the one where I'd grown up—and why did people there rarely even question their beliefs about drugs? A key problem is that poor people actually have few "competing reinforcers." Crack isn't really all that overwhelmingly good or superpowerfully reinforcing: it gained the popularity that it achieved in the hood (again, far less than advertised) because there weren't that many other affordable sources of pleasure and purpose and because many of the people at the highest risk had other preexisting mental illnesses that affected their choices.

And that was why, despite years of media-hyped predictions that crack's expansion across classes was imminent, it never "ravaged" the suburbs or took down significant percentages of middle- or upper-class youth. Though the real proportion of people who became addicted to crack in the inner city was low, it was definitely higher than it was among the middle classes, just as is true for other addictions, including alcohol. Money has a way of insulating people from consequences. In addition, it carries with it more reasons for abstaining—there are things a high-socioeconomic-status person *has* to do that are incompatible with being intoxicated. Becoming an addict is tantamount to disavowing one's social niche.

High socioeconomic status provides more access to employment, and alternative sources of meaning, purpose, power, and pleasure, as well as better access to mental health care. The differences in the prevalence of crack problems are mainly related to economic opportunity, not special properties of cocaine. While drug use rates are pretty similar across classes (and often, actually lower among the poor), addiction—like most other illnesses—is not an equal-opportunity disorder. Like cancer and heart disease, it is concentrated in the poor, who have far less access to healthy diets and consistent medical care.

Moreover, research on alternative reinforcers has now shown repeatedly that they can be effective in changing addictive behavior. This kind of treatment is called contingency management (CM). The idea comes from basic behaviorism: our actions are governed to a large extent by what we are rewarded for in our environment. These cause-and-effect relationships where a reward is dependent (contingent) upon the person either doing or (in the case of drugs) not doing a particular behavior can be used to help change all types of habits.

In fact, part of the reason we wanted to compare the responses of crack users to vouchers for cash in our study, as opposed to vouchers for merchandise, was ultimately to understand what types of reinforcement would work best to aid recovery. There is now a whole body of literature showing that providing alternative reinforcers improves addiction treatment outcomes. It is far more effective than using punitive measures like incarceration, which often is less useful in the long run. Although while incarcerated many people stop or at least reduce their drug use, jail and prison themselves don't provide positive alternatives to replace drug habits. When heavy drug users return to their communities, they are not better equipped to find work and support themselves and their families; instead, having a criminal record and a gap in their résumé makes finding work even harder.

Reward-based CM treatments are sometimes controversial because they can be portrayed in the media as "paying addicts to stop using." Many people think it's unfair to those who "do the right thing" by not taking drugs to see drug users getting paid to behave the way they should behave anyway. Cash rewards are especially touchy, since the users could presumably simply buy drugs with the money.

But I see it differently, and here's why. Indeed, we've all probably observed how people respond to rewards in multiple areas of life. It's often seen most clearly in parenting: for example, if my sons want a new computer, I expect them to maintain a certain GPA. In most workplaces, if the boss offers a raise for achieving certain goals, employees will do their best to hit those targets. Because drug use is governed by the same principles that govern other behaviors, contingency management treatment uses these ideas to change addictive behavior.

Importantly, using alternative reinforcers in treatment doesn't make it more expensive, in part because it makes it more effective. When contingency management techniques are specifically applied not only to supporting recovery but also to developing skills that are in demand by employers, the costs are cut even further because the work itself produces value, not to mention reducing people's need for government benefits.

One study randomly assigned treatment-seeking cocaine users to either contingency management plus behavioral counseling or to a traditional twelve-step focused counseling treatment, which involves referring people to meetings of twelve-step groups like Alcoholics Anonymous and teaching them about the steps involved. Patients in the contingency management arm of the study received vouchers for merchandise whenever they had drug-free urines. Fifty-eight percent of participants in the contingency management group completed the twenty-four-week outpatient treatment—compared to just 11 percent in the twelve-step group. In terms of abstinence, 68 percent achieved at least eight weeks cocaine-free, versus just 11 percent in the twelve-step condition.[8] And after the rewards are stopped, people in CM are no more likely to relapse than other treatment graduates. Since more people complete treatment with CM, this makes for an overall reduction in relapse.

More than three dozen studies have now been conducted on contingency management, used in the treatment of opioid, cocaine, alcohol, and multiple-drug addiction.[9] They show that contingency management typically does better than treatment that does not use it—and that larger, faster rewards are more effective than smaller and less quickly received incentives. This, again, is exactly what research on other types of behavior would predict. Cash, as we showed, is more effective than merchandise as a reinforcer.

The most exciting CM research currently being conducted is work by Ken Silverman and his colleagues at Johns Hopkins University. They have developed what they call a "therapeutic workplace" in which CM is used to help train drug users for jobs in data entry. One study, for example, found that the therapeutic workplace nearly doubled abstinence rates from opioids and cocaine among pregnant and postpartum addicted women, from 33 percent to 59 percent in urine samples taken three times a week.[10] And Silverman's group has replicated these findings several times, in different populations of people with addictions.

While there are multiple benefits to this line of research, one of the most important is that participants' drug-taking behaviors are being replaced with real-world job skills. In this way, these programs ultimately pay for themselves by helping those who were formerly unemployable become productive workers. When alternative reinforcers are made available to those who formerly lacked them, drug problems can be overcome.

And in my own case, at Columbia in the summer of 1999 I finally reaped the reward I'd been seeking for so long: a faculty position job at an Ivy League university. I'd continued putting in long hours, studying my human participants as intently as I'd once watched my rats (though, thankfully, I didn't have to operate on the people). At the New York State Psychiatric Institute, in upper Manhattan, I would hole up in my office, analyzing data and thinking about my research. Although the cubicle-sized room had a window with a breathtaking view of the Hudson River, I kept the shade down: the only thing I wanted to see was my data or the research papers I was reading. By this point, I was studying the effects of marijuana and methamphetamine as well as crack cocaine, so I needed to familiarize myself with the literature on those drugs.

And since our participants lived on-site 24-7, that's pretty much when I was there, too, overseeing the lab assistants and making sure everything was going as it should. I liked getting to know the participants: it not only helped the experiments run more smoothly but also gave me insight into their world, which allowed me to do better science. I now try to minimize the extent to which theories or stereotypes influence my view of drug users, especially if they are standing before me and I can collect my own data.

My mentor, Marian, was intensely supportive, always letting me know how much progress I was making and keeping me abreast of where I stood in terms of getting a faculty position. She told me late in 1998 that after I'd finished the year, I'd be getting a letter offering me a job, which would start on July 1. I felt immense pride when she told me—and even more so when the letter actually arrived, bearing Columbia's official letterhead and saying, "We want you to join the faculty as an assistant professor of clinical neuroscience." Indeed, that was probably the proudest moment of my life, the moment when I knew that I might be able to make a career of this science thing.

I didn't know that less than a year later, my world would be thrown into turmoil again, when I discovered that I had fathered a son, who was now sixteen, when I myself had been sixteen, back home.

CHAPTER 14

Hitting Home

If the relationship of father to son could really be reduced to biology,
the whole earth would blaze with the glory of fathers and sons.

—JAMES BALDWIN

Standing outside the VFW hall in Hollywood, Florida, I heard a young man cursing loudly, saying what sounded like my name repeatedly amid the string of profanities. I had been talking to my younger brother Ray and some of my cousins. We were attending Grandmama's funeral reception. It was October 13, 2004.

I'd had many more professional successes since becoming an assistant professor at Columbia in 1999: I had been awarded a multimillion-dollar grant from the National Institute on Drug Abuse (NIDA), making me an independent researcher running my own lab. I'd published about two dozen papers and had been asked to join NIDA's African American Researchers and Scholars Work Group, which advises the agency's director on drug-related issues unique to black people. I was making good progress toward tenure.

But as I'd worked my way up in academia, I'd also grown ever further away from my family. To put it bluntly, my emotional growth had not paralleled my professional achievements. Indeed, in many ways, I wasn't emotionally much different from the child I'd been when I'd left home. When something went wrong in my relationships, my main way of coping was to ignore it, suppress my feelings about it, or simply cut myself off from the person or people involved. That's what I'd done with my family. Not surprisingly, they were hurt by what they saw as my snobbish behavior, seeing my refusal to spend much time with them as evidence that I thought I was above them or was embarrassed by the way they lived.

From my perspective, I didn't know how to comfortably reach across the experiential and intellectual gap that now separated us. I didn't have the emotional tools. Ever since I'd joined the air force, it had gotten harder and harder to negotiate the vast differences between my world and theirs. Each step in my education moved me only further away, through forces mainly beyond my control. The more I tried to negotiate the mainstream, the more time I spent primarily with white academics, the less I felt able to communicate easily with my family. The distance stymied me.

Also, I didn't want to admit, even to myself, that I was catching hell in the white world. Trying to learn their language and cultural norms was more difficult and exhausting than my macho exterior would allow me to concede. Frankly, I was getting my ass kicked and had no one to talk with about how to deal with it and simultaneously maintain my sense of my own blackness. In college, I had Jim Braye to mentor me, but even he never had to deal in White America as a black, dreadlocked academic/researcher with three gold teeth, working at an Ivy League university.

At work, there was no one to whom I related. At home, Robin did her best to help me deal, but there were things about the black American experience that were foreign to her as a white woman. I also kept many of my concerns quiet, in order to avoid hurting her feelings. For example, I felt I just couldn't tell her when I wanted to go to community events alone, knowing that black people self-censor around even the most down and well-meaning whites.

Robin also wasn't fully aware of how often I had to grin and bear it when I felt I had been screwed over because of a racial slight. I had been the lowest-paid postdoc within our group at Columbia, despite having done two previous postdocs, which should have given me some seniority. My wife couldn't understand why I wasn't visibly outraged at every instance of a slight. Of course, most black people know that if they responded to the majority of explicit and oblique insults they receive on a daily basis, they would not only be exhausted but would also be quickly labeled hypersensitive and therefore, be marginalized. Staying cool is the best defense.

Still, the fake smile and air of detachment all wear on you. There were days when I couldn't just keep it inside and move on. All whites were the enemy when I felt like that. To protect Robin, I didn't express this stuff out loud and tried to suppress even the thoughts and feelings I had about it, but that itself began to eat away at me. I felt trapped and constrained by all these conflicting demands. I couldn't help starting to resent her, even though I knew it wasn't her fault. I know she felt the effects of this struggle.

But when I went back home to Florida, I faced an entirely different set of challenges. I tried my best not to seem condescending. However, even the way I spoke now began to seem like an insult to my family and friends there. Having broadened my vocabulary and begun talking in the way that the mainstream considers grammatically correct, it became more difficult with each passing year to relax my pronunciation of words toward my childhood speech patterns.

Lord knows I tried to be fluent in both street and mainstream vernacular in order to avoid being viewed as a traitor. I tried to show that I could, as Wideman put it in his classic memoir *Brothers and Keepers*, "chase pussy . . . fight, talk trash, hoop with the best. . . ." But now my normal speech was no longer that of the streets of South Florida. I felt like a fraud trying to pronounce words in the way that I had when I was coming up. So, I would remain relatively silent in order not to be branded an impostor or worse. That too made it harder and harder for me to connect to my siblings and cousins.

I'd watch myself interacting with but not connecting to siblings and cousins with whom I had been through hell and back. As a child they had looked after me, had seen that I was safe, had given me pocket change. Now I didn't even speak their language. Despite having read books by black authors describing similar phenomena, I couldn't let go of my pride and say, "Hey bruh, sis, or coz, I'm struggling . . ." Instead I began avoiding them, and the years quickly passed. Brothers, sisters, and cousins were now grandparents, and my nieces and nephews were now mothers and fathers.

When I was sued over the paternity of my son Tobias, the rift that had been dealt with by slowly cutting contact became overt and acute. It was most pronounced with my sister Joyce, the one I'd been closest with as a child and the person who felt most strongly that I now thought I was "better than" the rest of the family. She was the sister who most expressed the hurt and pain of our separation. She also had very strong views about Tobias.

At first I had denied that it was even possible that he was my child—and I told everyone as much. I just couldn't believe it could be true. To make matters worse, Joyce insisted that he

A photo of my mom (kneeling) and siblings.
Kneeling, from left, Ray, Gary, and me.
Standing, from left, Joyce, Patricia,
Beverly, Brenda, and Jackie.

was my son, long before I could bring myself to accept it. She said she'd seen me with his mother, which I didn't think possible since we'd been together only that one time.

"Fuck Carl Hart," the young man in the parking lot outside the VFW said, now distinctly. I looked up from my conversation and saw a young dark-skinned brother with dreads, wear-ing long jean shorts and a T-shirt. He had multiple tattoos and several gold teeth. He didn't look like anyone I knew but seemed to be in his teens or early twenties.

"You fucking talking to me?" I said, getting ready to get into it. My brother Ray pulled me aside. We were at a funeral recep-tion, after all.

"That's Tobias," he said as he tried to calm me down. Ray suggested that I could probably understand why someone in his situation might be angry with me. I just stared. I had had no idea that he was going to be there. I'm sure he showed up because my mother and his maternal grandmother were friends and he'd somehow learned from them that I would be visiting. Naively, I hadn't even considered the possibility that he might attend Grandmama's funeral. Ray pulled me away and Tobias left. But that was my unfortunate first encounter with my son.

By that time, I had already been paying child support for three or four years. The paternity suit had been settled almost immediately after I'd gotten the DNA results. I still didn't feel any emotional or psychological connection with him and I hadn't had any contact with his mother, other than through the court papers. But I did feel a tremendous amount of guilt about my handling of the situation.

Tobias took matters into his own hands. The day after the funeral, he came over to my sister Brenda's house, where I was staying, in order to apologize for how he'd behaved. Now only slightly more prepared for the encounter, I began to talk with him, or shall I say, I began to watch myself listening to him talk. I felt as dissociated from myself in my dealings with him as I was with the rest of my family.

Tobias was twenty-one at the time and had brought his own son, who was just a toddler. I picked the little boy up and played with him, but it didn't sink in until later, when everyone began teasing me, that I was actually a grandfather and this was my grandson. Smiling and interacting with the little guy was a welcome distraction.

Meanwhile, Tobias and I tentatively approached each other, trying to figure out how to negotiate some kind of relationship. I did understand why he was angry; I knew that I had desperately wanted to spend more time with my own father when I was growing up. I imagined how I would have felt if Carl Sr. had denied even being my dad and didn't even want to meet me after he'd been forced to pay child support.

I didn't think I had the right to say much, so I listened and thought maybe I could learn something. I was surprised at how happy Tobias was simply to be speaking with me, despite my cautious demeanor. Perhaps I was a better actor than I thought. I learned that he had grown up to be homophobic and hardened and also that he clearly knew how to take care of himself in the world from which I had once come.

I did explain to him that I'd had no idea that he'd even been born: his mother and I barely spoke the night we'd spent together or immediately afterward, let alone communicated months later about her being pregnant as a result. First he responded to this defensively, saying, "Damn, you blaming my mom?" I backed off. I said we'd both been young and I didn't know what she was thinking. I didn't mean to blame her. Maybe she was scared, I suggested.

It was then that he told me that she'd told him that some other brother was his father, a guy she'd been seeing at one point when he was growing up. He'd also apparently been told at least once that his real father was dead, so he'd received a number of different, conflicting stories about his paternity.

I wasn't sure what to do with this information. The best I

could do was to say again that she and I had both been too young
and that he shouldn't be too hard on her. I changed the subject.

"So what are you doing work-wise?" I asked.

He said, "Shit, you know what I do."

I didn't quite get it. Maybe I didn't want to.

"I'm slinging," he said, meaning that he was a street-level
pharmacist. He seemed almost to be daring me to make something of it. I didn't know what he knew about my profession
or area of interest as a researcher, but I did know that he was
trying to tell me that he was strong and didn't need anything
from anyone. I asked a few questions to show him that I got
that, along the lines of "How's business? You making enough to
handle your responsibilities?" He nodded affirmatively.

And when there was an awkward pause, I found myself questioning him about his education and trying to emphasize the
importance of completing high school or getting a GED, though,
at some level, I knew this was only a bandage for what amounted
to a cancer by this point. I really was at a loss for words. I was
accustomed to helping people solve problems by teaching and
I was in that mind-set when we spoke, wanting to fix him and
make it all right. Of course, that wasn't possible: here he was,
a young, uneducated black man in a world that had no use for
him, a fate I'd only narrowly avoided myself.

But advice wasn't what he wanted from me then anyway, I
later recognized. All he wanted was to speak with his father,
to tell him about his hopes and dreams and life. He wanted me
to know that he was going to be a good father, that he was a
good person. He wanted that affirmation from the man who had
brought him into the world, just like I'd wanted from my own
father as a child.

Meanwhile, I was still struggling with the fact that he was
my son and that he actually was in the life I might have had

myself if I'd stayed in Miami. I kept looking at him, but I really didn't see any of myself there other than in his defiance. I sure recognized that angry swagger and desperate need for respect. I didn't want to, but I did.

And truthfully, I didn't really want to look very closely. At the time, I didn't want to think too much about the other path my life could have taken, and be forced again to contemplate the differences between where I was now and the person I'd been growing up. I was surrounded by the starkness of that difference every time I came home. Still, we did manage to leave the lines of communication open.

And as I got to know him, I thought about the alternative reinforcers my other sons have had available to them that Tobias had either not been exposed to or not found ways to experience. I realized, too, that meeting Tobias had been especially shocking in comparison with my first encounters with my two other sons. The birth of my son Damon had been one of the deepest, most joyous, and most memorable experiences of my life. And by the time my son Malakai arrived six years later, I felt that I was actually starting to get the hang of this dad thing.

Though both of their births had been peak life experiences for me, I'd realized as I changed diapers, chased toddlers, and— before I knew it—found myself watching them play basketball and wondering when they'd be able to outplay me, that it wasn't at all the biological bond that made a father. It was the care, the daily repetitive care. It was being there and learning with them, having a life together.

And so seeing Tobias had felt like a slap in the face. I felt as though I was being held responsible for a child I'd had no say in raising. I wanted to do the right thing but I couldn't help feeling cheated. All the learning he'd done, all the reinforcement and punishment and extinction training he'd received for the most formative years of his life had had nothing to do with me. I'd almost literally been an unwitting sperm donor and yet this child was blood. The differences between him and my other sons and the arc of my childhood and his confounded me. I couldn't help thinking about those differences as I slowly got to learn more about his life.

Although I can't know for sure, I do have some speculations about some of the important differences. Unlike me, my son Tobias had never seriously participated in organized sports or even much in street games. He hadn't known the pleasure of becoming skilled at something through practice and using the fruit of hard work to win public competitions. He hadn't had a father like me in his life or older sisters like mine who nurtured him when his mother was unable to. His mother had been even younger and less well informed than MH had been when I was born; he didn't know the real story about his father. He didn't have even the limited academic success I'd had with math in elementary school. In fact, he doesn't appear to have ever been engaged at all by education and he dropped out before completing high school.

Tobias didn't have a Big Mama who stressed the importance of getting that degree, nor did he have a dream like mine of athletic glory, which led me to enlist in the air force rather than face the humiliation of not playing at least college-level ball. He didn't get military training, nor did he have the opportunity to travel and see a world different from the one he knew in South Florida. He hadn't found mentors who could teach him about black history and consciousness, real men who could show him the way to find different values than getting the most pussy (and seeing women in that demeaning fashion) and having a name on the street. The gap between us felt even more vast than the one between me and my family in Miami. At least I had a shared history with them.

When I met him, Tobias had so little mainstream cultural capital that he described me to his friends as a "teacher." He didn't understand the difference in status between a high school teacher and a college professor, let alone the differences between being a tenured professor or non-tenure-track lecturer, or between being at an Ivy League or a less prestigious school. Just as I'd been as a teenager, he was completely isolated from the mainstream.

I didn't know how to reach him or provide him with appropriate and helpful alternative reinforcers. He isn't a drug addict; he's a young black man with no high school diploma and limited employable skills in a country that sees him as a problem, not a resource. The unemployment rate for black men at the end of 2012 was about 14 percent, twice the number for white men.[1] Those problems don't have answers in the neuropsychopharmacology that I study.

I began to realize that I would need to speak out if I was to ensure that my work didn't lead people to the wrong conclusions about drugs and the causes of social problems.

CHAPTER 15

The New Crack

There are in fact two things, science and opinion; the former begets knowledge, the latter ignorance.

—HIPPOCRATES

One afternoon in mid-2005 I received a phone call from the U.S. drug czar's office, the ONDCP, a component of the Executive Office of the President. Initially, I thought, Oh shit, I must be in trouble! But that wasn't it. They were phoning to request my participation in a roundtable discussion about the drug methamphetamine. The purpose of this roundtable, the caller explained, was to educate writers about the real effects of methamphetamine so that stories written about the drug would be more authentic. The writer participants would consist of individuals who wrote for a variety of magazines and television shows. I happily agreed to take part because this seemed to be a departure from previous "educational" efforts by ONDCP. These were the same folks who in the late 1980s brought us the public service announcement (PSA), "This is your brain on drugs." During the spot, a man holds up an egg and says, "This is your brain." Then he picks up a frying pan and says, "This is drugs." Then he cracks open the egg, fries the contents, and says,

"This is your brain on drugs." Finally, he asks, "Any questions?" This PSA is one of the most ridiculed antidrug advertisements of all time because of its simplistic and inaccurate portrayal of drug effects.[1]

Today, ONDCP's slogan is "Relying on science, research and evidence to improve public health and safety in America." So, perhaps one of the goals of the roundtable, I figured, was to provide the writers with information with foundations in evidence rather than fear-based anecdotes. In addition to me, the panelists were an assistant U.S. attorney, an undercover narcotics agent, and a methamphetamine "addict." Because I was one of the few scientists studying the effects of methamphetamine in people, my role was to summarize the current state of our scientific knowledge about the drug. I began by saying that methamphetamine is approved by the U.S. Food and Drug Administration to treat ADHD and narcolepsy. The attendees were surprised. How could this awful drug that they had heard about be sanctioned for anything? Then I presented data from my studies showing that methamphetamine produced the same effects as the better-known prescription medication Adderall (generic name: a mixture of amphetamine salts). The chemical structure of the two drugs is nearly identical (see figure 2).

This too was surprising to most in the room. Like amphetamine, methamphetamine increases energy and enhances one's ability to focus and concentrate; it also reduces subjective feelings of tiredness and cognitive disruptions brought about by fatigue and/or sleep deprivation. Both drugs can increase blood pressure and the rate at which the heart beats. I explained that several nations' militaries, including our own, have used (and continue to use) amphetamines since World War II because of these properties.[2] The drug helps soldiers fight better and longer.

My fellow panelists were horrified because my lecture was in

*Figure 2. Chemical structure of amphetamine (active
ingredient in Adderall), left, and methamphetamine.*

stark contrast to the stories they told about methamphetamine.
The attorney showed a slide presentation filled with disheveled
children of alleged illegal methamphetamine makers. "These are
America's children," she asserted, hoping to evoke a sympathetic
emotional response. Her remarks were echoed by the narcotics agent, who declared that methamphetamine was unlike any
drug he had seen in his twenty years of service. They asserted
that the drug produced an addiction more severe than any other
drug, including crack cocaine. The police officer further warned
that users of methamphetamine are so violent that Taser guns
are ineffective at stopping them. "These people are animals," he
said, and insisted that more intensive methods are necessary for
stopping someone high on methamphetamine. The officer concluded his remarks with an anecdote so ghastly that the audience
moaned in unison. He stated that methamphetamine causes cognitive impairment so severe that it can lead parents to decapitate
their own children; he swore he had witnessed this firsthand.

Judging from the audience's responses, the anecdotes were
effective. They urgently wanted know why law enforcement
hasn't done more do to get this awful drug off the streets. Or
how could anyone in their right mind take such a destructive
chemical? None of the writers raised questions about the verac-

ity of the stories told by the attorney or the narcotics agent, even though they had just heard conflicting information about the drug. The world was flat again. My mind raced with thoughts about that 1914 *New York Times* article describing "Negro cocaine fiends" and how southern police forces had to exchange their revolvers for heavier-caliber weapons because cocaine gave black people superhuman strength. I was baffled that others in the room didn't recognize how myths about drugs are recycled from one generation to another; I was disappointed because I believed this roundtable would be different. I thought evidence from science would inform our view on the drug. Instead the roundtable turned out to be similar to other drug discussions sponsored by the government, an exchange in hysteria and ignorance. I was also angry because I knew such hysteria unfairly vilified methamphetamine users and decreased their willingness to seek help if needed.

The discussion also reminded me of the exaggerated claims about crack cocaine two decades before. As I pointed out earlier, that drug was believed to be so powerfully addictive that even first-time users would become addicted. It had also been linked to the deaths of two promising young athletes—Len Bias and Don Rogers—although later it became clear that the athletes had taken large amounts of powder and not crack cocaine. Powder cocaine was seen as a recreational drug for the wealthy.

Few people asked whether the sentencing disparity between the two forms of cocaine was based on scientific evidence. In 1986, there were only two scientific publications on smoked cocaine. Both studies contained a number of limitations, which decreased their relevance in public policy discussions. As a result, the law that created the 100:1 crack-powder sentencing ratio was based entirely on anecdotal reports. This in itself isn't necessarily a bad thing as long as lawmakers understood the limits of this approach and were prepared to alter the law as new, more complete knowledge dictated.

By the early 1990s, concern about the dangers of crack intensified and lots of money was pumped into the war on that drug. Not only were law enforcement budgets increased but more money was also allocated for research. Now scientists had a stake in the crack hysteria game. As a result, the scientific database on crack cocaine grew substantially within a few years. As I stated earlier, the data showed that both forms of cocaine produce identical effects; these effects are predictable. That is, as the dose is increased, so are the effects, whether they are blood pressure and heart rate *or* subjective "high" and addictive potential. The evidence clearly indicated that the 100:1 ratio exaggerated the harms associated with crack and that the sentencing disparity was not scientifically justified. To punish crack users more harshly than powder users is analogous to punishing those who are caught smoking marijuana more harshly than those caught eating marijuana-laced brownies.

At the same time, some began raising concerns that crack-powder laws disproportionately targeted blacks. Congress directed the U.S. Sentencing Commission to issue a report examining the federal cocaine laws. The commission is the federal agency responsible for, among other tasks, reducing unwarranted sentencing disparities. In February 1995, it issued its report. The report examined pharmacology, the ways the drug is taken, societal impacts, cocaine distribution and marketing, cocaine-related violence and crime, the legislative history of cocaine penalties and constitutional challenges, and data related to federal drug offenses. It was thorough. It found that nearly 90 percent of those sentenced for crack cocaine offenses were black, even though the majority of users of the drug were white. This conflicted with most people's perception because news reports and popular media almost always showed black crack smokers. As a result of these findings, the commission submitted to Congress an amendment to the sentencing guidelines that would have equalized penalties for powder and crack cocaine offenses, that is, the crack-powder ratio would have gone from 100:1 to 1:1. Congress passed and President Bill Clinton signed legislation disapproving the guideline amendment. In a statement Clinton explained the rationale for his decision to block the amendment: "We have to send a constant message to our children that drugs are illegal, drugs are dangerous, drugs may cost you your life—and the penalties for dealing drugs are severe." He continued: "I am not going to let anyone who peddles drugs get the idea that the cost of doing business is going down." Subsequent reports and recommendations by the commission in 1997, 2002, and 2007 were equally unsuccessful in bringing about meaningful changes to the cocaine laws.

Many prominent individuals criticized the unwillingness of lawmakers to eliminate the cocaine sentencing disparity. In 1997, Michael S. Gelacak, then vice chairman of the Sentencing Commission, wrote, "Congress and the Sentencing Commission have a responsibility to establish fair sentencing standards that protect the public. . . . We have jointly failed in our approach toward crack cocaine sentences, and the result is seriously disparate sentences. We should not lose sight of that overriding reality. . . . The only real solution to the injustice is to eliminate it." Ten years later, even presidential candidate Barack Obama had added his voice to the growing chorus of criticism: "[L]et's not make the punishment for crack cocaine that much more severe than the punishment for powder cocaine when the real difference between the two is the skin color of the people using them. Judges think that's wrong. Republicans think that's wrong. Democrats think that's wrong, and yet it's been approved by Republican and Democratic Presidents because no one has been willing to brave the politics and make it right. That will end when I am President."[3] On August 3, 2010, President Obama signed legislation that decreased, but did not eliminate, the sentencing disparity between crack and powder cocaine offenses. The new law reduced the sentencing disparity from 100:1 to 18:1.

Some celebrated this change as a significant step toward ending a historic wrong. I am not one of them. In 1964, when asked whether the United States had made sufficient progress toward racial equality, Malcolm X said, "If you stick a knife in my back nine inches and pull it out six inches, there is no progress. . . . The progress is healing the wound." Accordingly, I think the sentencing differences should be completely eliminated because there is no scientific justification for the differential treatment of crack and powder cocaine under the law. This seems the ethical thing to do in light of the evidence and ONDCP's claim to rely on science and evidence.

I sat there in the methamphetamine roundtable and wondered whether the same mistakes would be made with this drug as were made with crack cocaine. There certainly were plenty of signs suggesting this. Like with crack cocaine in the mid-1980s, a relatively small number of individuals from a derided group were seen as users of methamphetamine. They were white but gay, poor, or rural. In 2005, about a half million people reported that they had used methamphetamine in the past thirty days (an indication of "current use"). This number is small when compared with the 15 million people who smoked marijuana within the same period. Whenever a "new" drug is introduced to a society and a relatively small number of marginalized individuals use that drug, incredible stories about the drug's effect can be told and accepted as fact. This is so because few people have the experience with the drug to challenge questionable claims.

We saw this in the 1930s when authorities said that marijuana caused people to become psychotic and commit murder. These claims were often unchallenged and taken as fact. In fact, they were a major reason that the federal law (Marihuana Tax Act of 1937) essentially banning marijuana was passed. At the time, marijuana use was confined to a small number of minorities and "hipsters." Of course, today, if an individual says that marijuana use leads to insanity and murder, he or she would not be taken seriously.

Another similarity with the "crack scare" of the 1980s was the increasing number of stories written about methamphetamine in the national press. On August 8, 2005, *Newsweek* ran a dramatic cover story called "The Meth Epidemic." Use of this drug, according to the magazine, had reached epidemic proportion. The evidence suggested otherwise. At the height of methamphetamine's popularity, there were never more than a million current users of the drug. This number is considerably lower than the 2.5 million cocaine users, the 4.4 million illegal prescription opioid users, or the 15 million marijuana smokers during the same period. The number of methamphetamine users has never come close to exceeding the number of users of these other drugs.[4]

Coverage was filled with accounts of desperate users turning to crime to support their use of the "dangerously addictive" drug. Many articles focused on the "littlest victims." The *New York Times* headlined one story, DRUG SCOURGE CREATES ITS OWN FORM OF ORPHAN, describing an apparent rise in related foster care admissions and reports of addicted biological parents who were impossible to rehabilitate. The paper quoted a police captain who said methamphetamine "makes crack look like child's play, both in terms of what it does to the body and how hard it is to get off."[5] The paper also claimed, "Because users are so highly sexualized, the children are often exposed to pornography or sexual abuse, or watch their mothers prostitute themselves."[6] Attorney General Alberto Gonzales called it "the most dangerous drug in America," and President George W. Bush proclaimed November 30, 2006, National Methamphet-amine Awareness Day. Back in 1986, President Ronald Reagan proclaimed the entire month of October Crack-Cocaine Aware-ness Month. The parallels were frightening.

At the end of the ONDCP discussion, we were asked to meet with writers in small groups to answer any lingering questions. Dozens lined up to meet with the police officer and attor-ney. They wanted to hear more about how methamphetamine caused gay men to engage in sexual practices that increased HIV rates; how it kept people up for consecutive days without sleep-ing; how the drug made people behave irrationally; and how it ruined people's teeth and made them unattractive. While some of the writers were undoubtedly there simply seeking a sexy story to sell, I think most genuinely wanted to learn about the drug and, if needed, to warn the public about its dangers. They weren't thinking about separating anecdote from evidence. They had just heard from a U.S. attorney and a cop that this drug was nasty stuff. The government invited both of these individuals as experts on the topic. As a result, there didn't seem to be a need to separate fact from fiction. Of course the information was fac-tual. Otherwise, it would not have been presented in a forum sanctioned by the government, would it?

I pondered this and other questions as I rode the subway back to my lab. Why was my data so inconsistent with the stories told by the other panelists? Was I out of touch with the way people use drugs in the real world? Maybe the doses that I tested were too low, I thought. I had intentionally started off with low doses to ensure the safety of my research participants. At that point, the largest dose I had given was 20 mg, which is considerably lower than doses reportedly used by methamphetamine addicts. Perhaps the individuals described by the prosecutor and police officer used much larger doses than those tested in my studies. This might explain our disparate conclusions. I also thought about how methamphetamine is typically used outside the lab— snorted, injected, or smoked. This ensures that the drug hits the brain more quickly and produces more intense effects. In my studies, it was swallowed. When taken this way it produces the least intense effects. Given these caveats, I questioned whether the data collected in my studies was relevant to the situation in the real world. I figured the hysteria about methamphetamine had to reflect something about reality and that my studies, up until that point, had not captured it.

Over the next seven years, I went about trying to resolve this issue. I searched the human literature to see if anyone had studied larger methamphetamine doses when the drug was snorted, smoked, or injected. There was virtually nothing. I thought about José Martí's famous quote in his 1882 essay "On Oscar Wilde": "A knowledge of different literatures is the best way to free one's self from the tyranny of any of them." So I read the literature on animal studies looking for information that might be relevant to human addiction. These studies showed that the drug caused extensive damage to certain brain cells and produced severe learning and memory problems. Aha, I got it! Finally, here was some data that was in line with popular anecdotes about methamphetamine. But as I looked more closely it became clear that the animal results had serious limitations and might not be applicable to people. For one, the amounts of methamphetamine given to animals are far more than amounts taken by methamphetamine addicts. If one gave similarly high doses of caffeine or nicotine to animals, the same serious toxic effects would be seen.

But when animals were given methamphetamine doses comparable to those used by people, the destructive effects were not observed. During my graduate education, the notion that methamphetamine damaged brain cells was an unquestioned fundamental truth in drug research. Now this basic belief needed to be qualified, making it difficult to extrapolate to people.

Next, I studied the literature on the long-term effects of methamphetamine in addicts. These were people who had used the drugs for many years. In these studies, abstinent methamphetamine addicts and a control group (usually non–drug users) completed a comprehensive set of cognitive tests over the course of several hours, and the results were compared to determine whether the cognitive functioning of the methamphetamine addicts was normal. Of course, normality is a relative concept that is determined by not only comparing performance of the methamphetamine group with the performance of a control group but also comparing the methamphetamine group's scores with those from a normative dataset, taking into consideration the individual's age and level of education. These requirements are important because they allow us to take into account the relative contribution of age and education in terms of the individual's score and adjust the score accordingly. Simply stated, it would be inappropriate to compare the vocabulary scores of a sixteen-year-old high school dropout with those of a twenty-two-year-old college graduate. The older college graduate would be expected to outperform the younger dropout.

Study after study found that methamphetamine addicts had severe cognitive impairment. In one study by Sara Simon and colleagues, the apparent impairments were so bad that it led them to warn: "The national campaign against drugs should incorporate information about the cognitive deficits associated with methamphetamine. . . . Law enforcement officers and treatment

With members of my lab at an end-of-year celebration.

providers should be aware that impairments in memory and in the ability to manipulate information and change points of view (set) underlie comprehension . . . methamphetamine abusers will not only have difficulty with inferences . . . but that they also may have comprehension deficits . . . the cognitive impairment associated with [methamphetamine abuse] should be publicized. . . ."[7] As I read this and similar papers more critically, I noticed something intriguing. While it was true that the controls had outperformed methamphetamine addicts on a few tests, the performance of the two groups wasn't different on the majority of tests. More important, when I compared the cognitive scores of the methamphetamine addicts in the Simon study against scores in a larger normative dataset, none of the methamphetamine users' scores were outside the normal range.[8] This meant that the cognitive functioning of the methamphetamine users was normal. This should have tempered the researchers' conclusions and prevented them from stating such dire warnings. What's more, the methamphetamine literature was filled with similar unwarranted conclusions; as a result, the apparent methamphetamine addiction–cognitive impairment link has been widely publicized—numerous articles have appeared in scientific journals and the popular press.

The reporting of brain imaging findings has been especially misleading. On July 20, 2004, for example, the *New York Times* printed an article titled, THIS IS YOUR BRAIN ON METH: A "FOREST FIRE" OF DAMAGE. It stated, "People who do not want to wait for old age to shrink their brains and bring on memory loss now have a quicker alternative—abuse methamphetamine . . . and watch the brain cells vanish into the night." This conclusion was based on a study that used magnetic resonance imaging (MRI) to compare brain sizes of methamphetamine addicts with non-drug-using healthy people.[9] The researchers also looked at the correlation between memory performance and several brain structural sizes. They found that methamphetamine users' right cingulate gyrus and hippocampus were smaller than those of controls by 11 and 8 percent, respectively. Memory performance on only *one* of four tests was correlated with hippocampal size (that is, individuals with larger hippocampal volume performed better). As a result, the researchers concluded, "chronic methamphetamine abuse causes a selective pattern of cerebral deterioration that contributes to impaired memory performance." This interpretation, as well as the one printed in the *Times* article, is inappropriate for several reasons.

First, brain images were collected at only one time point for both groups of participants. This makes it virtually impossible to determine whether methamphetamine use caused "cerebral deterioration," because there might have been differences between the groups even before methamphetamine was ever used. Second, the non–drug users had significantly higher levels of education than methamphetamine users (15.2 versus 12.8 years, respectively); it is well established that higher levels of education lead to better memory performance. Third, there were no data comparing methamphetamine users with controls on any memory task. This, in itself, precludes the researchers from making statements regarding impaired memory performance caused by methamphetamine. Nonetheless, the only statistically significant cognitive finding was a *correlation* of hippocampal volume and performance on one of the four tasks. This finding is the basis for the claim that methamphetamine users had memory impairments, because the hippocampus is known to play a role in some long-term memory; but other brain areas are also involved in processing long-term memory. The size of these other areas was not different between the groups. Finally, the importance on everyday functioning of the brain differences is unclear because an 11 percent difference between individuals, for example, is most likely within the normal range of brain structure sizes.

This example is not unique. The brain imaging literature is replete with a general tendency to characterize any brain differences as dysfunction caused by methamphetamine (as well as other drugs), even if differences are within the normal range of human variability.[10] It would be like comparing the brains of police officers who have less education with those of college professors who have obtained a PhD, and concluding that the officers are cognitively impaired as a result of any differences that might be noted. This simplistic thinking is the main thrust behind the notion that drug addiction is a brain disease. It certainly isn't a brain disease like Parkinson's disease or Alzheimer's disease. In the case of these illnesses, one can look at the brains of affected individuals and make pretty good predictions about the illness involved. We are nowhere near being able to distinguish the brain of a drug addict from that of a non–drug addict.

Because the literature wasn't as informative as I'd hoped, I wrote and received a grant to study larger methamphetamine doses in individuals snorting the drug. These laboratory studies detailed the immediate and short-term effects of the drug on measures of cognitive functioning, mood, sleep, blood pressure, heart rate, and the drug's addictiveness. I tested doses up to 50 mg, which were, at the time, the largest doses tested in people. All of the drug doses were given in a double-blind manner— the research participants didn't know whether they were getting placebo or real methamphetamine, nor did the medical staff monitoring the sessions. The research participants were carefully selected to make sure they were in excellent medical condition. All were addicted to methamphetamine and used more than 100 mg of the drug on a weekly basis. I wanted to make sure that I was not exposing them to more drug in the lab than they used outside the lab. Similar to the cocaine studies I had previously conducted, we intentionally recruited people who were not seeking treatment, because we felt it was unethical to give methamphetamine to someone trying to stop using.

In the first experiment, we simply had research participants snort one dose of methamphetamine while our medical team carefully monitored their vital signs for twenty-four hours. We also asked the participants to do cognitive tests and rate their mood before and several hours after the drug was given. The findings were consistent with data from our previous studies when we gave the drug by mouth.[11] Participants reported feeling more euphoric and their cognitive functioning was improved. These effects lasted about four hours. The drug also caused significant increases in blood pressure (BP) and heart rate that lasted for up to twenty-four hours. The maximum levels were about 150/90 (BP) and 100 (beats per minute). While these elevations were undoubtedly significant, they were well below levels obtained when most people are engaged in a vigorous activity such as physical exercise. Another finding was that the drug reduced the amount of time our participants slept.[12] For example, when they took placebo, participants got about eight hours of sleep on that evening. But when the 50 mg dose was given, they got only six hours of sleep. Together, the results indicated that a large snorted dose of methamphetamine produced expected effects. The drug didn't keep people up for consecutive days, it didn't dangerously elevate their vital signs, nor did it impair their judgment. Around the same time, other researchers were studying the drug when it was injected or smoked and they were getting similar results.[13]

The human laboratory data were at odds with anecdotal reports and conventional wisdom. Maybe I hadn't asked the right question. One of the most popular beliefs about methamphetamine is that it is highly addictive, more so than any other drug. In the next set of experiments, I set out to address this issue. Under one condition, I gave methamphetamine addicts a choice between taking a big hit of methamphetamine (50 mg) or five dollars in cash. They took the drug on about half of the opportunities. But when I increased the amount of money to twenty dollars, they almost never chose the drug.[14] I had gotten similar results with crack cocaine addicts in an earlier study.[15] This told me that the addictive potential of methamphetamine was not as had been claimed; its addictiveness wasn't extraordinary. My results also showed me that methamphetamine addicts, just like crack addicts, can and do make rational decisions, even when faced with a choice to take the drug or not. This was consistent with the literature assessing cognitive functioning of methamphetamine users, but as noted above, only if you looked carefully.[16]

Still, the popular view of methamphetamine remained unchanged. Most media portrayals continued to emphasize unrealistic effects and exaggerate the harms associated with the drug. For example, in January 2010, NPR ran a story titled, "This Is Your Face on Meth, Kids." The story described a California sheriff who was trying to stop young people from experimenting with methamphetamine. With the help of a programmer, he developed a computer program that digitally altered teenagers' faces to show them what they would look like after using methamphetamine for six, twelve, and thirty-six months. These young people watched their images change from those of healthy, vibrant individuals to faces marred by open scabs, droopy skin, and hair loss. They were told that these were the physiological effects of using methamphetamine. Ninety percent of individuals who tried methamphetamine once, they were also told, would become "addicted." How could such inaccurate information be given to naive students, let alone be reported on NPR, I thought.

There is no empirical evidence to support the claim that methamphetamine causes one to become physically unattractive. Of course, there have been the pictures of unattractive methamphetamine users in media accounts about how the drug is ravaging some rural town. You may have also seen the infamous "meth mouth" images (extreme tooth decay). But consider this: methamphetamine and Adderall are essentially the same drug. Both drugs restrict salivary flow, leading to xerostomia (dry mouth), one proposed mechanism of meth mouth. Adderall and generic versions are used daily and frequently prescribed—each year they are among the top one hundred most prescribed drugs in the United States—yet there are no published reports of unattractiveness or dental problems associated with their use.

The physical changes that occurred in the dramatic depictions of individuals before and after their methamphetamine use are more likely related to poor sleep habits, poor dental hygiene, poor nutrition and dietary practices, and media sensationalism. With regard to the addictiveness of methamphetamine, the best available information clearly shows that the majority of people who try methamphetamine will not become addicted.[17]

The media and general public were not the only ones caught up in the methamphetamine hysteria. Many scientists were also bamboozled. From 2006 to 2010, I was a member of an NIH grant review committee. The committee was composed of about forty scientists with diverse expertise. One of our main tasks was to evaluate the scientific merits of research grant proposals submitted by drug abuse scientists. We frequently reviewed proposals seeking funds to study methamphetamine. Many of the proposals argued that the drug produced brain damage, while others pointed to the cognitive impairments caused by methamphetamine. They seemed to have accepted, as a foregone conclusion, that any use of this drug was destructive. These arguments were compelling to some on the review committee. The problem was that they were not supported by evidence; instead they were misrepresentations of the data. I am not suggesting that the scientists who wrote the grants did this intentionally. I don't think they did. I do believe, however, that the scientists understood quite well the mission of their proposed funder—National Institute on Drug Abuse (NIDA)—and this understanding shaped their grant proposal.

NIDA's mission "is to lead the Nation in bringing the power of science to bear on *drug abuse and addiction*." Drug abuse and addiction are only limited and negative aspects of the many effects produced by drugs. Of course, drugs like methamphetamine produce other effects, including positive ones such as improved cognitive performance and mood, but that isn't a part of NIDA's mission. And scientists seeking research money from NIDA are well aware that they must emphasize the negative effects of drugs in order to get funded. Upton Sinclair's famous quote aptly describes this situation: "It is difficult to get a man to understand something when his salary depends upon his not understanding it."[18] Consider also that NIDA funds more than 90 percent of all research on the major drugs of abuse. This means that the overwhelming majority of information on drugs published in the scientific literature, textbooks, and popular press is biased toward the negative aspects of drug use.

I am not suggesting that the negative consequences of drug use shouldn't be the focus of research funded by NIDA. Focusing on the pathological aspects of drug use is extremely important for developing effective treatments for drug addiction. But the current disproportionate focus on the bad effects of drugs tends to leave us with a skewed perspective. It has helped to create an environment where certain drugs are deemed evil and any use of these drugs is considered pathological. As I have repeatedly pointed out throughout this book, most people who use any drug do so without problems. This is not an endorsement for the legalization of drugs. It's just a fact. The near-exclusive focus on the negative effects of drugs has also contributed to a situation where there is an unwarranted and unrealistic goal of eliminating certain types of drug use at any cost. Too often marginalized groups absorb the bulk of the cost. It has been well documented that certain minority communities have been particularly affected by our zeal to get rid of certain drugs. The human cost of this misguided approach is incalculable, as hundreds of thousands of men and women, including my own family members, languish in prison as a result.

In an effort to draw attention to the misinterpretations that plague the methamphetamine scientific literature, I wrote a critical review article that assessed more than fifty peer-reviewed research studies on the short- and long-term effects of the drug on brain and cognitive functioning.[19] I concluded that meth-amphetamine addicts were overwhelmingly within the normal range on both measures. But, despite this fact, there seems to be a propensity to interpret any cognitive and/or brain difference(s) as clinically significant abnormalities.

Before any paper is published in a scientific journal, experts in the field must review it anonymously. These reviews are often brutal. They sometimes question your intellectual capacity to be employed as a scientist. So, when I received the reviews for my paper I was expecting harsh criticism because I was essentially calling into question an entire body of research. To my sur-prise, the reviewers' comments were extremely laudatory: "This review is comprehensive and extremely well written. Dr. Hart and colleagues certainly challenge the status quo and should be applauded for writing a provocative paper and taking what will surely be characterized as an unpopular position. . . . The general message is somewhat of a wake-up call to the field. . . ." It's too soon to know the exact impact the paper will have on the field, but shortly after its publication, *Scientific American* highlighted it in an article that questioned whether the methamphetamine hysteria is limiting the availability of effective medicines.[20]

This all got me thinking even more about the consequences of presenting biased, exaggerated, or misleading drug information to the public. As an educator, I worried that we would lose cred-ibility with many young people, and that as a result they would reject other drug-related information from "official" sources, even when the information was accurate. Undoubtedly, this has contributed to numerous preventable drug-related accidents. I thought about the distorted claims made about crack cocaine and how this led to egregious racial discrimination. During the "crack

Presenting my research findings at a scientific meeting.

era," I didn't know any better. I was ignorant. Ignorance could not be used as an excuse in the current case of methamphetamine. I knew better. I had published my research findings in some of the finest science journals and had coauthored one of the bestselling drug textbooks. And each semester, my Drugs and Behavior course was one of the most popular undergraduate classes at Columbia. Still, voices like mine were rarely included in national discussions on drug education or public policies about drugs. My voice was not included because I had serious trepidations about exposing myself in this way. I knew that some would say that I had an agenda, implying that I might be less than objective. This is one of the worst criticisms that can be leveled against a scientist. Others would attempt to label me as reckless by distorting my views as advocating for the complete legalization of drugs.

Ultimately, it became clear to me that I had to speak out beyond the walls of the academy. I started giving lectures in local community centers, at the YMCA, student-organized conferences, bars and cafés, at museums, or anywhere else I was asked to speak. I spoke with high school students and their parents about the real effects of drugs and ways to decrease associated harms. I lectured at other universities about the foolish way the country deals with drugs and the biases I was starting to see in the questions we asked about drugs in science.

A frequent question from parents was "What about the children? Isn't it better to exaggerate drug-related harms so we keep our children away from them?" Black people rarely asked this question; it almost always came from white parents. I tried to be as patient as possible in my response. I'd point out that I too was a concerned parent with three sons—two within the critical ages of concern—and how I have educated the two that I raised about drugs without exaggerations. I explained that my twenty-plus years of drug research experience has taught me many important lessons, but perhaps none more important than this—drug effects are predictable. As you increase the drug dose, there is more potential for toxic effects. Black boys' and men's interactions with the police, however, are not predictable. I worried all the time about the very real possibility that my own children would be targeted by law enforcement because they "fit the description" of a drug user or because someone thought they were under the influence of drugs. Too often in these cases the black youngster ends up dead. Ramarley Graham and Trayvon Martin both were believed to have drugs or be under the influence of them.

In addition to giving more public talks, I was invited to join nonscientific advocacy organizations. I was particularly intrigued by an invitation to join the board of directors of America's leading organization dedicated to changing the drug laws: the Drug Policy Alliance (DPA). This was a difficult decision. I knew it would put me in an awkward position with the man who'd recruited me to Columbia, Herb Kleber. Herb had served under President George H. W. Bush as deputy drug czar from 1989 to 1991; many of his views were in line with those of most politicians who claim that drugs are evil and that we should pursue a "drug-free America" at any cost. The DPA was about as far away on the political spectrum from Herb's view of drug policy as it was possible to get. When I told Herb that I was considering joining the DPA board, he warned that it wasn't a wise thing to do at this stage of my career—I was being considered for tenure. In order to reach a decision, I also spoke with a prominent black former DPA board member, who told me to be careful about being used because of my race. In this person's view, DPA was a white organization concerned mainly with legalizing marijuana so that white boys could smoke without fear of harassment by the police. As a result, they were far less concerned about fighting the racial discrimination so prevalent in the drug war. I considered all of these things but ultimately joined the board. It was my way of making a very public statement about my views on our country's misguided drug policies that disproportionately target blacks. It was also my way of making sure that the leading group challenging current drug policies was fully informed about and had access to the best scientific research.

One of DPA's mottos is that it promotes "alternatives to current drug policy that are grounded in *science*, compassion, health and human rights." This really appealed to me because it suggested that this organization understood the importance of using science to inform drug policies and ultimately enhance health and human rights. After I had spent five years of service on the DPA board, however, it became inescapably obvious that their understanding of science was slightly different from my own. I naively thought that the scientific evidence would guide DPA's focus and positions taken much as it does in my own research. In my view, if DPA had simply followed the scientific evidence, their priorities would look quite different. Rather than a predominant focus on marijuana legalization and increasing the number of states with medical marijuana programs, unbiased scientifically informed public education about drugs would be the major priority. The evidence that I have presented throughout this book suggests that the average person is woefully ignorant about illegal drugs and their use. As a result, an organization like DPA could fill an important knowledge gap if they spearheaded public education campaigns aimed at enhancing the intellectual tone around drug-related issues that substantially impact public health. For example, because the majority of heroin overdoses occurs in combination with another sedative—mostly alcohol—a massive media campaign warning users to avoid combining heroin with other sedatives would not only be educational—it could also be life-saving. I also recognize that government agencies, such as ONDCP and NIDA, should take the lead on these efforts, but they have consistently demonstrated their unwillingness and/or inability to do so.

I discovered, though, that DPA faced the same pressures and limitations that many other nonprofits face—the donors influence priorities. That is why, for the past several years, marijuana law reform has been the major focus of DPA, even while it has played a major role in bringing attention to racist stop-and-frisk laws in New York City. As it turns out, like ONDCP, the use of the term *science* in DPA's slogan seems to be a matter of convenience rather than a commitment to truth that guides the organization's positions and focus. Of course, this cunning use of language is more egregious when employed by ONDCP, because it's a government agency and not an advocacy organization. These sad realizations contributed to me being more vocal on the board and writing this book in an effort to educate the public about drugs.

CHAPTER 16

In Search of Salvation

If the society today allows wrongs to go unchallenged, the impression is created that those wrongs have the approval of the majority.

—BARBARA JORDAN

GOD is offering Salvation, apply through Jesus Christ," read a huge billboard on Sunrise Boulevard. Sitting in rush hour traffic, I reflected on where I'd just been. It left me demoralized and definitely in need of some salvation, even though I'm not particularly religious. As part of my research for this book, I was interviewing relatives and old friends in South Florida and had just spent the last hour with my cousin Louie. Growing up, he and I had shared a bed at Big Mama's; he was the math whiz that I admired. Now he lived in a halfway house just off the Florida Turnpike in Fort Lauderdale, and it had been nearly thirty years since I saw him last.

"What's up man, you know who I am?" I asked the skinny man standing in front of me. He was wearing a wife-beater T-shirt and oversize blue jeans. The home care attendant had pointed out Louie, who was standing outside speaking with another resident. "Big Jun," he responded. When we were boys Louie had always called me Lil' Carl or Junior; now I was Big Jun. I was surprised that he'd even recognized me, because my appearance had changed so much over the past three decades. His had, too. He stood about six feet but weighed, at most, 110 pounds. His face was so emaciated that you could see nearly every bone. The few remaining teeth he had looked like they were on the way out. I was shocked, disturbed, and profoundly sad but showed only that I was happy to see him because I didn't want to hurt his feelings. Over the years, I had become a master at masking my emotions, although this skill had been seriously tested in the course of writing this book.

We slapped hands, did the bro-hug thing, and without interruption, Louie talked for the next hour. He talked about the various crimes he had committed over the years and the amounts of money he had stolen and stashed. I learned that the police had beaten him up on many occasions and that he believed his insides had been replaced with those from other people. He contemplated whether he had done the right thing by not becoming a police informant, "I didn't tell nothing. Maybe I should've start telling. I did plea guilty and didn't snitch on nobody. They wouldn't let me go home since I didn't give 'em no information. I should've turned state on 'em."

Louie's thoughts were disjointed and difficult to follow. He jumped from one subject to another without a break or transition and paced around the small yard the entire time I was there. His involuntary, repetitive movements were a textbook case of tardive dyskinesia brought on by taking antipsychotic medications for more than two decades. Although the details aren't clear, family lore has it that he was initially put on these medications in the ER after having a "bad reaction" to an unknown street drug. And when he was sent to prison, they kept him on them in order to keep him obedient and calm—a chemical strait-jacket.

In graduate school, I'd learned quite a bit about antipsychotics and what they were used for. These were the drugs used to treat schizophrenia and related illnesses. The simplistic idea is that psychotic behaviors such as those seen in schizophrenia are caused by overactivation of dopamine cells in the brain. Antipsychotic drugs block dopamine receptors and thereby prevent excessive dopamine activity. Behaviorally, these drugs quiet the voices in the heads of schizophrenics and reduce their paranoia and agitation. The problem is that the older generation of these medications, the type that Louie was prescribed, block dopamine receptors so extensively that the brain compensates by increasing the density of dopamine receptors. The brain is now hypersensitive to dopamine, and after years of treatment, the person develops tardive dyskinesia and becomes even more susceptible to psychotic symptoms. In other words, the treatment for psychotic symptoms can actually cause these symptoms. It's a trap.

With each passing minute, Louie's voice became background noise and I felt more and more grief and despair. I wondered how this could have happened but already knew the answer, because his story wasn't unique. I had seen similar scenarios with other male loved ones. Virtually all had been initially caught up in the system via a drug charge while in their teens and early twenties, which began a vicious cycle from which they couldn't escape. What's worse is that the cycle wasn't even new. One hundred years ago, on September 29, 1913, the *New York Times* printed an article that described how a white mob in Mississippi lynched and shot two black young men, one eighteen and the other twenty, because they were suspected of starting "a reign of terror" under the influence of cocaine. The following day the paper reported that the town's two thousand black residents had been forced to walk past the bullet-riddled bodies of the two boys to view them; this, the article continued, "had a remarkably quieting effect on the negro population." I would imagine it did.

Of course, we no longer lynch people for violating drug laws. Today the damage is far less visible and starts more subtly. The educational and vocational skills that sustain people throughout life are usually obtained during young adulthood, from the late teens throughout the twenties. This is a critical period. I, for example, spent most of my young adult years in classrooms and labs learning how to think and write. These skills have allowed me to support my family financially, which gives me a sense of worth and manhood. As a result, I have a stake in this society and do my best to make a contribution to it. It doesn't matter whether the contribution is in paying taxes or doing public service or takes some other form. The point is that society and I both benefit from me having a stake in it.

In contrast, so many of the black boys with whom I grew up don't have any stake in our society. They didn't acquire the necessary skills and didn't get the needed support during that critical period. Instead they were under the supervision of a system that doesn't seem to understand or care about the importance of black men being invested in this society. Supporters of this system have an irrational focus on eliminating certain drugs and are preoccupied with those who violate drug laws, especially if they are black. Selective enforcement of drug laws seems to serve as a tool to marginalize black men and keep them in the vicious cycle of incarceration and isolation from mainstream society. I am not arguing that people shouldn't be sanctioned for legal infractions. There are many cases in which sanctions are appropriate. However, the penalty should not be so severe that the penalized young person is unable to recover and stake a claim in society. In such cases, we all lose. The young person's loss is obvious. The general public is deprived of the contribution that would have been made if the person were a stakeholder. With no real stake in the larger society, many of my friends and relatives feel they have nothing to lose. And as James Baldwin observed, "The most dangerous creation of any society is the man who has nothing to lose."[1]

After speaking with my sisters, I could see that we were beginning to lose some of my nephews. They had already started to repeat the incarceration-isolation cycle. What could I tell them? Hell, I don't even know what to tell my own son Tobias. He has spent time behind bars for a drug violation and doesn't have a high school diploma; nor does he have an employment track record or any prospects for a legitimate job. I'd recently spoken with him on a previous visit and he caught me up on the current events in his life. I learned more about baby-mama drama than I cared to know. "Man, they always want some shit," he complained about the difficulties of dealing with the three different mothers of his children. At the same time, he was extremely proud to be a father five times. It was his badge of honor, something that "real" men do, even though he was unemployed. And unless there is some radical change in this society, his chances of getting a legitimate job are extremely bleak because these irrefutable facts remain: he is a black male who has a drug conviction and limited marketable skills. Like Louie, he too is trapped.

Don Habibi, my old mentor from the University of North Carolina Wilmington, was fond of saying, "Once you know, you cannot not know." There was a period in my life when I was unaware of the forces preventing Tobias and people like him from legitimately competing in mainstream society. That time has passed; I have come to understand that the game is fixed against them. That's why I am frequently disheartened and stressed when asked what to tell someone in Tobias's position. I recognize that I can't give up on him or our society. So when we met last, I again encouraged him to get his GED and a legitimate job. I told him about my brother Gary, who had also dropped out of high school and dabbled in cocaine sales, but would eventually graduate college and own a multimillion-dollar company. I didn't tell him that Gary had never been convicted of a crime, nor did I tell him that Gary had only one child when he started to turn his life around. That contextualization might have been too daunting. After all, I was trying to convince Tobias, as well as myself, that he too could do it.

Along Gary's journey, I had given him a copy of Nathan McCall's *Makes Me Wanna Holler*. It was the first book that he had ever read cover to cover. He found it inspirational. So I bought Tobias a copy, too, and asked him to read it so we could discuss it. I also got him Bob Marley's *Survival* CD, printed out the lyrics, and asked him to listen to it with a particular focus on the track "Ambush in the Night." I explained that the song poignantly describes how the system is stacked against people like him and how sometimes it's nice to know that someone else gets it. Still, this felt insufficient for what he faced. It felt like giving a Band-Aid to a gunshot wound victim who is profusely bleeding when everyone knows that a surgeon is needed to remove the bullet so the healing can begin.

A redeeming aspect of writing this book was that it afforded me an opportunity to mend family relationships that had been damaged by years of unspoken words and distance. On several occasions I met separately with MH and Carl and got to know them as people and not just parents. From MH, I'm sure I got my twisted sense of humor. She'd frequently poke fun at her grandchildren: "Malik wants to be thug and don't know how to be one. He ain't even man enough to pee straight. He better sit his light-in-the-behind-ass down." She made me laugh constantly when we got together. Another thing that she did was to help keep me connected to people from my past. "You remember Lil' Mama?" she'd ask. Invariably, I'd say no. MH would continue: "She told me to tell you hello and to remind you that she saved you from getting many ass-whippings." "Oh yeah, now I remember her, Lil' Mama," I'd reply.

My interactions with Carl were equally rewarding but centered primarily on sports. He wanted to make sure that I continued to support the Miami-based professional teams. "What do you think of those Heat?" I didn't have the heart to tell him that I've never been a Heat supporter. The Miami Heat joined the NBA for the 1988–1989 season, four years after I had left the area. So I never developed an emotional bond with that team as I had with the Dolphins. Nonetheless, it's clear to me that Carl spurred my interests in athletics, and were it not for athletics, this book probably would have never been written. My participation in high school athletics required that I maintain a minimum GPA, which ensured that I would graduate. Carl and I reminisced about the time when we went to see the Muhammad Ali–George Foreman fight, 1974's "Rumble in the Jungle," on closed-circuit television at the convention center. It was a special night; it was our birthday. I also learned that he speaks with Tobias on a regular basis, offering guidance and support, and that he hasn't had a drink in nearly twenty years.

As I spent time with my parents, I couldn't help thinking about my own young children and the time that I wasn't spending with them. Damon was now eighteen and preparing to go off to college and Malakai was six years younger, attending a middle school that charges tuition rates comparable to a college. The environment in which Robin and I are raising them is utterly different from the one in which I was raised. This is a source of anxiety and relief. I sometimes worry that we have pampered them too much. Would they be able to fend for themselves should something happen to Robin and me? My siblings and I joke about how MH made it clear to us that we were on our own very early in life, especially if we got into trouble with the law. One of her favorite lines was "If you go to jail, don't call me." MH firmly believes that her child-rearing philosophy is the reason for her children's success in life. Her children, however, have a different perspective.

Robin and I have been fortunate to shield our children from the traps that face so many other black boys, including Tobias and my nephews. Damon and Malakai don't seem to have the emotional scars that I carried from my childhood. They are thoughtful and verbally expressive, even when emotional. Both have participated in athletics and the arts since they were very young. Each has already read more books than I had upon completion of my undergraduate studies; for them, an undergraduate education is the minimum expectation. They have traveled throughout the United States and have been to foreign countries. Importantly, they are staking their claim in this society. The thing that pleases me most, however, is that they are happy and cheerful. Much of their free time is spent together playing games, laughing and joking. When watching Damon and Malakai interact, I am often reminded of the time when Louie and I were kids climbing the huge sapodilla tree in Big Mama's yard. "Don't go too high," Louie would say. Because he was older, he felt compelled to look after me and make sure I didn't step on a weak branch and fall.

After saying good-bye to Louie, I sat in the car and cried, because I felt as though I had failed to look after him as he had

Billboard.

done for me when we were kids. Prior to writing this book, I hadn't cried since I was a child. Now, in the car, a flood of tears poured from my eyes. I thought about all of the other Louies we've failed to look after. I thought about all the years that I spent away from my Florida family in order to obtain an education that seems inadequate to help solve the problems they face. The tears continued streaming down as I thought about the tremendous promise that Louie once showed; I felt crushed that we both couldn't have been scientists. After several minutes, I gathered myself and started the car. Johnny Cash was on the radio singing, "There will be peace in the valley for me, dear Lord I pray. . . ." And I slowly drove away.

CHAPTER 17

Drug Policy Based on Fact, Not Fiction

It's time for America to get right.

—FANNIE LOU HAMER

So, are you saying that we should legalize hard drugs like cocaine, heroin, and methamphetamine?" The question was in response to the presentation that I had just given to a group of white, aging New York City hipsters. They were fairly well educated; you know, the NPR type. Some were even professionals such as neurologists, psychologists, and social workers. All had come to a basement bar in Brooklyn to hear me speak at their monthly "secret science club" meeting.

The room was dimly lit, smelled of alcohol, and filled to capacity—several would-be attendees were turned away. Bodies rubbed up against bodies like we were in a popular dance club. Some even reeked of marijuana smoke. And as I stood on the brightly lit stage—so bright that I had to wear sunglasses—I couldn't help thinking back to my childhood when I was a DJ spinning records in similar settings, except then, the audiences were all black. "I mean I totally agree that war on drugs has been a huge failure. And I even support legalizing marijuana, but I'm *not* in favor of legalizing the hard drugs," the attractive thirty-something woman wearing a Public Enemy T-shirt continued.

I wasn't surprised by her question and comments. It wasn't the first time that one of my presentations had been met with skepticism or incredulity. And to be fair, I had just told this audience, many of whom took great pride in their critical thinking skills, that they had been hoodwinked and miseducated for much of their lives about what drugs do and don't do. I used a mountain of scientific data to call into question some of the purported damaging effects of the "hard drugs" on brain functioning. I explained that there has been an ongoing concerted effort to overstate the dangers of drugs like cocaine, heroin, and methamphetamine. The primary players in this effort are scientists, law enforcement officials, politicians, and the media.

While I acknowledged the potential for abuse and harm caused by these drugs, I emphasized that there had been extensive misinterpretation of the scientific evidence and considerable hyping of anecdotal reports. This situation, I explained, has not only wrongly stigmatized drug users and abusers; it has also led to misguided policy making. Does this mean that drug legalization is the only option available to us when we reconsider what drug policies should be in place? Of course not. Drug prohibition, the most prevalent current form of drug policy, and drug legalization are on opposite ends of the drug policy continuum. There are multiple options in between.

One such option is drug decriminalization. Decriminalization is often confused with legalization. They are not the same thing. Here's the major difference: Under legalization, the sale, acquisition, use, and possession of drugs are legal. Our current policies regulating alcohol and tobacco, for those of legal age, are examples of drug legalization. Under decriminalization, on the other hand, the acquisition, use, and possession of drugs can be punished by a citation much like traffic violations are. Mind you, drugs still are *not* legal, but infractions do not lead to criminal convictions—the one thing that has prevented so many from obtaining employment, housing, governmental benefits, treatment, and so on. This is crucial when you consider this fact: each year, more than 80 percent of arrests in the United States for drug offenses involves only simple possession.[1] Sales of all illicit drugs, however, remain criminal offenses under decriminalization laws.

Drug decriminalization isn't a new concept. In fact, a handful of states, including California and Massachusetts, have decriminalized marijuana. Although the specifics vary from state to state, in general the laws read something like this: any person caught with less than an ounce of marijuana or smoking in public is punishable by a civil fine of one hundred dollars. No state has decriminalized other illegal drugs. You might ask, why not? Well, before answering, it might be informative for us to look at the Portugal experiment.

Back in 2001, Portugal took the unprecedented step of decriminalizing all illegal drugs. That's right, cocaine; heroin; methamphetamine; 3,4-methylenedioxymethamphetamine (MDMA, aka ecstasy and molly); everything. Here's how it works there. Acquisition, possession, and use of recreational drugs for personal use—defined as quantities up to a ten-day supply—are no longer criminal offenses. Users stopped by police and found to have drugs are given the equivalent of a traffic ticket, rather than being arrested and stigmatized with a criminal record. The ticket requires them to appear before a local panel called (in translation) the Commission for Dissuasion of Drug Addiction, typically consisting of a social worker, a medical professional like a psychologist or psychiatrist, and a lawyer. Note that a police officer is not included.

The panel is set up to address a potential health problem. The idea is to encourage users to honestly discuss their drug use with people who will serve as health experts and advisers, not adversaries. The person sits at a table with the panel. If he or she is not thought to have a drug problem, nothing further is usually required, other than payment of a fine. Treatment is recommended for those who are found to have drug problems—and referral for appropriate care is made. Still, treatment attendance is not mandatory. Repeat offenders, however—fewer than 10 percent of those seen every year—can receive noncriminal punishments like suspension of their driver's license or being banned from a specific neighborhood known for drug sales.

How has decriminalization been working out for the people of Portugal? Overall, they have increased spending on prevention and treatment, and decreased spending for criminal prosecution and imprisonment. The number of drug-induced deaths has dropped, as have overall rates of drug use, especially among young people (15–24 years old). In general, drug use rates in Portugal are similar, or slightly better, than in other European Union countries.[2] In other words, Portugal's experiment with decriminalization has been moderately successful. No, it didn't stop all illegal drug use. That would have been an unrealistic expectation. Portuguese continue to get high, just like their contemporaries and all human societies before them. But they don't seem to have the problem of stigmatizing, marginalizing, and incarcerating substantial proportions of their citizens for minor drug violations. Together, these are some of the reasons that I think decriminalization should be discussed as a potential option in the United States.

"So, why isn't decriminalization of all illegal drugs given

serious consideration in this country?" yelled a deceptively aged man standing in the center of the room. The salt-and-pepper hair and creases in his face suggested he was in his late forties or early fifties, but his skinny jeans and Chuck Taylor Converse sneakers implied he was much younger. I replied, "Of course, the answer to this question will vary depending upon who is being asked to address it. And consideration of all of the possible answers is beyond the scope of this talk." In the preceding pages, however, I have tried to provide the reader with information that can allow one to address this question in a more critical manner. Briefly, we're too afraid of these drugs and of what we think they do. Our current drug policies are based largely on fiction and misinformation. Pharmacology—or actual drug effects—plays less of a role when policies are devised. As such, we have been bamboozled to believe that cocaine, heroin, methamphetamine, or some other drug du jour is so dangerous that any possession or use of it should not be tolerated and deserves to be severely punished. Decriminalization is inconsistent with this misguided perspective.

In order to begin a serious national discussion about decriminalization, first, the public will have to be reeducated about drugs, separating the real potential dangers from monstrous or salacious fable. While I hope this book is a significant step in that direction, others (for example, scientists and public health officials) will also be needed in our reeducation efforts. And given how entrenched some drug myths are, one should not expect change to occur within a short period. This will engender considerable disappointment and frustration. I am reminded of the words of my dear friend Ira Glasser, former director of the American Civil Liberties Union, when he was asked how long it will take for us to see meaningful drug policy reform. Ira responded, "The fight for justice is not a sprint . . . it's a marathon relay race. You can't see where the track ends. You can just take the baton and run as hard as you can and as fast as you can and as far as you can. . . ."

Ira's comments also remind us that reeducation of the public about drugs will take a team effort. For one, scientists who study illegal drugs can be extremely helpful in this process. But you should also know that scientists are not all equal in their ability to think critically and rationally about drugs. For example, a researcher who studies the neurotoxic effects (causing damage to brain cells) of MDMA in rodents is not necessarily the best person to educate the public about that drug's effects on people. In their experiments, these researchers typically inject very large amounts of the drug several times a day, for consecutive days. In many experiments, the animal is given as much as ten times the amount of drug that a human would take. So it wouldn't be surprising that MDMA, given in these large doses, can cause damage to brain cells. What is surprising, however, is that some scientists, on the basis of these results, communicate dire warnings to the public that MDMA should not be used even once because it causes brain damage. With teammates like this, you don't need opponents. I assure you that if you administered similar excessively large alcohol or nicotine doses to animals, you would observe similar or even more toxic effects. But these findings are probably not relevant to human drug use because we take considerably smaller amounts of drug.

Given the vast amount of conflicting information, I recognize that it can be difficult trying to determine who is a credible drug expert. In your attempt to evaluate the drug information presented, it might be helpful to ask a few simple questions: (1) How much drug was given to the animals and is it similar to amounts used by people? (2) Was the drug injected or swallowed and do people use the drug in this way? (3) Were the animals first given smaller amounts of the drug to allow the development of tolerance, which prevents many toxic effects, or were naive animals just given larger amounts initially? (4) Were the animals housed in isolation or in groups? All of these factors potentially influence drug effects on the brain and behavior. You should be skeptical when "experts" attempt to extrapolate data collected in laboratory animals to humans without appropriate consideration of these critical factors.

Law enforcement is another profession that is frequently called upon to educate the public about drugs. Few efforts have had a more harmful effect on public education and health. In general, police officers are trained to apprehend criminals and prevent and detect crime, in the service of maintaining public order. They don't receive training in pharmacology, nor do they receive any in psychology or any other behavioral science. As I have consistently pointed out within these pages, the effects of drugs on human behavior and physiology are determined by a complex interaction between the individual drug user and her or his environment. Without the appropriate training it's extremely difficult to draw conclusions about how a particular drug might have influenced someone's behavior.

It's true that cops make numerous drug arrests, but it would be a mistake to assume that they become knowledgeable about the effects of drugs as a result. Being pursued or arrested by the police is an aberrant situation. This, in itself, even without any drugs, can cause heightened suspicion, anxiety, fear, and the fight-or-flight response in the suspect. Consider also that some people who are arrested for drugs have preexisting psychiatric disorders, while others may be intoxicated from using multiple drugs, including alcohol. When all of these complexities are added to an already abnormal setting, it is often difficult to tease apart the effects of a particular drug from those of non-drug influences. Yet, in some public education campaigns carried out by police officers, the disturbing behaviors are uncritically attributed to certain illegal drug effects. This is an important vehicle through which drug-related myths are perpetuated. The point is that law enforcement officials are not qualified to serve as drug education experts simply because they make arrests that may involve drugs.

Both the scientists who study toxicity in animals and the police who arrest users and sellers often have a limited view of the complexity of the ideas I have presented to you. No one whose professional experience focuses only on one aspect of illicit drug use can be considered a real expert in the sense of being able to imagine all the intended and unintended consequences of continuing our current policy of treating illicit drug use primarily as a criminal issue.

The media, too, is another major source of drug misinformation. Throughout this book I have provided multiple examples of how the media has generally fanned the flames of drug hysteria. It seems as though there's a "new deadly drug" nearly every year. And invariably some police officer or politician is interviewed, warning parents about the dangers this drug poses to their children. (Of course, neither cop nor elected official should be the professional educating the public about the potential effects of drugs.) Usually, after the hysteria has subsided, we discover that the drug in question wasn't as dangerous as we were initially told. In fact, it wasn't even new. But by then the new laws have been passed and they require stiff penalties for possession and distribution of the so-called new, dangerous drug. I am not optimistic that the media will change its reporting on drugs anytime soon. Drug stories are just too sexy, and sex sells everything from newspapers to documentary films.

Nonetheless, you should know that scientists have studied nearly all of the popular recreational drugs in people. We have learned a great deal about the conditions under which either positive or negative effects are more likely to occur. Unfortunately, this knowledge is rarely disseminated to the public, primarily because of the irrational belief that it might lead one to engage in drug use. In light of the fact that there are already more than 20 million Americans who use illegal drugs regularly, it seems that a rational approach—one that aims to reduce drug-related harms—would be to share what we've learned with drug users and those in positions to help keep them safe. Otherwise, it seems that we do society a major disservice.

If more people were aware of a few simple facts that we've learned, this would substantially enhance public health and safety. First, inexperienced drug users should be discouraged from taking drugs in the manner in which experienced users do. Experienced users tend to take drugs in ways that get them to the brain quickly, that is, by smoking or intravenous injection. Because smoking and shooting drugs intravenously produce more potent effects, the likelihood of harmful consequences is increased with these methods. Alternatively, taking a drug by mouth is usually safer than other ways of consuming drugs for two reasons: (1) the stomach can be pumped in case of an overdose, which isn't possible with smoked or injection overdoses; and (2) some of the drug will be broken down before reaching the brain, resulting in fewer drug effects.

Second, healthy sleep habits should be stressed for all drug users because prolonged sleep loss can cause deterioration of mental functioning. In severe cases, even without drugs, hallucinations and paranoia may also occur. Because amphetamines and cocaine reliably reduce fatigue and offset performance decrements, some may repeatedly take these drugs to lessen problems associated with sleep loss. This is a less than ideal approach. One of the most consistent effects of stimulants is the disruption of sleep, which means that repeated use can exacerbate problems related to sleep loss. Given the vital role that sleep plays in healthy functioning, regular users of stimulants should be mindful of their sleep durations and avoid drug use near bedtime.

Finally, certain drug combinations should be avoided because they increase the risk of overdose. The heroin-alcohol combination and oxycodone-diazepam combination are two popular examples. While it is theoretically possible to die from an overdose of any of these drugs alone, in practical terms this is extremely rare. But each year there are several thousand deaths in the United States in which the use of these combinations is mentioned. In virtually every overdose death involving an opioid, for example, some other substance is present. Most of the time it is alcohol. One should always take care when combining two drugs or any drug with alcohol.

"Thank you for your time and thoughtful questions and comments," I said to the secret-science clubbers as I prepared to exit the stage. But before I could leave, dozens of people lined up and huddled around me. Some had pressing follow-up questions, while others wanted to share their stories, seek advice, or just say thanks. It was reminiscent of a time when I used to watch my DJ mentor and brother-in-law, Richard "Silky Slim," skillfully convey to each person that his or her opinion mattered. Unfortunately, Silk didn't go on to become the well-known entertainer that we all thought he would. Instead he was convicted of a drug charge and served more than a decade in federal prison.

Since his release, he and I have spoken at great length about his experiences with the justice system and the unfairness of our current drug policies. To be honest, his story is one that provides motivation for me to do my best to change the way we regulate the illegal drugs discussed in this book. Whenever I write something relevant on the topic, I send him a copy. Here's a recent text message I received from him in response to an op-ed that I published calling for the elimination of the sentencing disparities between powder and crack cocaine:[3] "Hey brother Carl, Great fucking piece you wrote man. That made my heart start racing again just thinking about the injustice I had to endure. Thanks man, that was beautiful! God Bless." I sincerely hope that my efforts help to prevent many of the policy mistakes that were made in the past.

ACKNOWLEDGMENTS

would like to thank two people who helped nurture this book along a path that proved to be more difficult than I initially thought: Claire Wachtel and Maia Szalavitz. Claire went far beyond her editorial responsibilities by serving in multiple roles. Thank you for treating me like a writer and for being my sounding board, clinical psychologist, and friend. Without your subtle but clear and firm guidance, this book would have been a superficial and uninteresting read. Maia, your professionalism is unmatched. You kept me on schedule despite my best efforts to delay and avoid dealing with difficult personal issues within these pages. I am also deeply appreciative of your teaching me how to write an engaging story. This is not a lesson taught in most science education programs.

Melissa and Marc Gerald, of course minus your efforts, this project would not have been undertaken. While serving on an NIH grant review committee with Melissa, late one evening at dinner, she suggested that I meet with her literary agent brother about writing a book. I thought she was just being kind by indulging my atypical ideas. As it turned out (and I am grateful), Marc agreed with her and put the Agency Group's best efforts into seeing the project through to completion. Sasha Raskin, my coagent, thanks for being patient with my endless inquiries about the publishing process.

On most days, I feel fortunate to have an intellectual home at Columbia University in the departments of psychology and psychiatry and the Institute for Research in African-American Studies, where I learn from some of the most talented thinkers. I owe a tremendous debt to my many coauthors, colleagues, and students. These individuals took the time to teach me about drugs, science, and life. The arguments and discussions in which we engaged helped to shape several of the ideas put forth in this book. I am particularly indebted to Charles Ksir, James Rose, Fredrick Harris, Robert Krauss, Norma Graham, Lynn Paltrow, Rae Silver, Catalina Saldaña, and Susie Swithers. Some of these individuals even read and reacted to early drafts of the manuscript.

To my family, thank you all for your support and allowing me to share your stories. Robin's early encouragement provided much of the fuel that helped me power through some inevitably difficult portions of the process. Writing this book would have been impossible were it not for the sharp memories of Jackie, Brenda, Beverly, Patricia, Joyce, Gary, and Ray. In addition, Ray's ability to find obscure newspaper articles about Carver Ranches and our childhood friends is truly amazing. His research helped me to tell a richer story.

Finally, I would be remiss if I did not acknowledge a few government programs for their contributions toward my physical and intellectual development without which this book might not have been written: Aid to Families with Dependent

Children (welfare as we once knew it), the National Institute on Drug Abuse's Supplemental Grant for Minorities in Biomedical and Behavioral Research, and the National Institute of Mental Health–Society for Neuroscience Predoctoral Minority Fellowship. In recent years, programs aimed at redressing past American racial discrimination have come under attack. Without such programs, however, I seriously doubt that I would have become the scientist, educator, and tax-paying citizen that I am today.

NOTES

CHAPTER 1: WHERE I COME FROM

1. J. C. Anthony, L. A. Warner, and R. C. Kessler, "Comparative Epidemiology of Dependence on Tobacco, Alcohol, Controlled Substances, and Inhalants: Basic Findings from the National Comorbidity Survey," *Experimental and Clinical Psychopharmacology* 2 (1994): 244–68; L. A. Warner et al., "Prevalence and Correlates of Drug Use and Dependence in the United States. Results from the National Comorbidity Survey," *Archives of General Psychiatry* 52, no. 3 (March 1995): 219–29; M. S. O'Brien and J. C. Anthony, "Extra-Medical Stimulant Dependence Among Recent Initiates," *Drug and Alcohol Dependence* 104 (2009): 147–55; Substance Abuse and Mental Health Services Administration, *Results from the 2011 National Survey on Drug Use and Health: Summary of National Findings*, NSDUH series H-44, HHS publication no. (SMA) 12-4713 (Rockville, MD: Substance Abuse and Mental Health Services Administration, 2012).

2. Gwendolyn Mink, *Poverty in the United States: An Encyclopedia of History, Politics, and Policy* (Santa Barbara, CA: ABC-CLIO, 2004), vol. 1, p. 187.

3. Linda Swanson, "Racial/Ethnic Minorities in Rural Areas: Progress and Stagnation," U.S. Department of Agriculture Economic Research Service, AER731 (August 1996), www.ers.usda.gov/publications/aer731/aer731g.pdf. Also Manning Marable, *How Capitalism Underdeveloped Black America* (London: Pluto Press, 1983), p. 45.

4. Substance Abuse and Mental Health Services Administration, Office of Applied Studies, *Results from the 2004 National Survey on Drug Use and Health: National Findings*, DHHS publication no. SMA 05-4062, NSDUH series H-28, 2005, http://www.oas.samhsa.gov/p0000016.htm#2k4.

5. Thomas P. Bonczar, "Prevalence of Imprisonment in the U.S. Population, 1974–2001," U.S. Department of Justice, Bureau of Justice Statistics Special Report, NCJ 197976 (August 2003), www.policyalmanac.org/crime/archive/prisoners_in_US_pop.pdf.

CHAPTER 2: BEFORE AND AFTER

1. B. A. Pan, M. L. Rowe, J. D. Singer, and C. E. Snow, "Maternal Correlates of Growth in Toddler Vocabulary Production in Low-income Families," *Child Development* 76, no. 4 (July–August 2005): 763–82, http://www.ncbi.nlm.nih.gov/pubmed/18300434; M. L. Rowe, "Child-Directed Speech: Relation to Socioeconomic Status, Knowledge of Child Development and Child Vocabulary Skill," *Journal of Child Language* 35, no. 1 (February 2008): 185–205, http://www.ncbi.nlm.nih.gov/pubmed/16026495; M. L. Rowe and S. Goldin-Meadow, "Differences in Early Gesture Explain SES Disparities in Child Vocabulary Size at School Entry," *Science* 323 (February 2009): 951–53, http://www.ncbi.nlm.nih.gov/pubmed/19213922.

2. P. K. Piff et al., "Having Less, Giving More: The Influence of Social Class on Prosocial Behavior," *Journal of Personality and Social Psychology* 99, no. 5 (November 2010): 771–84; M. W. Kraus, S. Côté, and D. Keltner, "Social Class, Contextualism, and Empathic Accuracy," *Psychological Science* 21, no. 11 (November 2010): 1716–23.

CHAPTER 3: BIG MAMA

1. D. K. Ginther et al., "Race, Ethnicity, and NIH Research Awards," *Science* 333 (2011): 1015–9.

2. C. M. Mueller and C. S. Dweck, "Praise for Intelligence Can Undermine Children's Motivation and Performance," *Journal of Personality and Social Psychology* 75, no. 1 (July 1998): 33–52.

CHAPTER 4: SEX EDUCATION

1. R. A. Wise, "The Neurobiology of Craving: Implications for the Understanding and Treatment of Addiction," *Journal of Abnormal Psychology* 97 (1988): 118–32; G. F. Koob, "Drugs of

Abuse: Anatomy, Pharmacology and Function of Reward Pathways," *Trends Pharmacological Sciences* 13 (1992): 177–84.

2. J. Olds and P. Milner, "Positive Reinforcement Produced by Electrical Stimulation of the Septal Area and Other Regions of Rat Brain," *Journal of Comparative and Physiological Psychology* 46 (1954): 419–27.

3. C. Hart and C. Ksir, "Nicotine Effects on Dopamine Clearance in Rat Nucleus Accumbens," *Journal of Neurochemistry* 66 (1996): 216–21; C. Ksir et al., "Nicotine Enhances Dopamine Clearance in Rat Nucleus Accumbens," *Progress in Neuro-Psychopharmacology and Biological Psychiatry* 19 (1995): 151–56.

4. W. A. Cass et al., "Differences in Dopamine Clearance and Diffusion in Rat Striatum and Nucleus Accumbens Following Systemic Cocaine Administration," *Journal of Neurochemistry* 59 (1992): 259–66.

5. J. Zhu et al., "Nicotine Increases Dopamine Clearance in Medial Prefrontal Cortex in Rats Raised in an Enriched Environment," *Journal of Neurochemistry* 103 (2007): 2575–88; J. Zhu, M. T. Bardo, and L. P. Dwoskin, "Distinct Effects of Enriched Environment on Dopamine Clearance in Nucleus Accumbens Shell and Core Following Systemic Nicotine Administration," *Synapse* 67 (2013): 57–67.

6. G. F. Koob, "Drugs of Abuse: Anatomy, Pharmacology and Function of Reward Pathways."

7. L. Hechtman and B. Greenfield, "Long-Term Use of Stimulants in Children with Attention Deficit Hyperactivity Disorder: Safety, Efficacy, and Long-Term Outcome," *Paediatric Drugs* 5, no. 12 (2003): 787–94.

CHAPTER 5: RAP AND REWARDS

1. H. R. White and M. E. Bates, "Cessation from Cocaine Use," *Addiction* 90, no. 7 (July 1995): 947–57.

2. A. J. Heinz et al., "Marriage and Relationship Closeness as Predictors of Cocaine and Heroin Use," *Addictive Behaviors* 34, no. 3 (March 2009): 258–63.

3. M. D. Resnick et al., "Protecting Adolescents from Harm: Findings from the National Longitudinal Study on Adolescent Health," *Journal of the American Medical Association* 278, no. 10 (1997): 823–32.

4. B. K. Alexander, R. B. Coambs, and P. F. Hadaway, "The Effect of Housing and Gender on Morphine Self-Administration in Rats," *Psychopharmacology* 58 (1978): 175–79; P. F. Hadaway et al., "The Effect of Housing and Gender on Preference for Morphine-Sucrose Solutions in Rats," *Psychopharmacology* 66 (1979): 87–91.

5. C. Chauvet et al., "Effects of Environmental Enrichment on the Incubation of Cocaine Craving," *Neuropharmacology* 63 (2012): 635–41; M. D. Puhl et al., "Environmental Enrichment Protects Against the Acquisition of Cocaine Self-Administration in Adult Male Rats, but Does Not Eliminate Avoidance of a Drug-Associated Saccharin Cue," *Behavioural Pharmacology* 23 (2012): 43–53; D. J. Stairs, E. D. Klein, and M. T. Bardo, "Effects of Environmental Enrichment on Extinction and Reinstatement of Amphetamine Self-Administration and Sucrose-Maintained Responding," *Behavioural Pharmacology* 17 (2006): 597–604.

6. M. E. Carroll, S. T. Lac, and S. L. Nygaard, "A Concurrently Available Nondrug Reinforcer Prevents the Acquisition or Decreases the Maintenance of Cocaine-Reinforced Behavior," *Psychopharmacology* (Berlin) 97, no. 1 (1989): 23–29.

7. M. Lenoir et al., "Intense Sweetness Surpasses Cocaine Reward," *PLoS One* 2, no. 8 (August 2007): e698.

8. M. A. Nader and W. L. Woolverton, "Effects of Increasing the Magnitude of an Alternative Reinforcer on Drug Choice in a Discrete-Trials Choice Procedure," *Psychopharmacology* (Berlin) 105, no. 2 (1991): 169–74.

9. S. T. Higgins, W. K. Bickel, and J. R. Hughes, "Influence of an Alternative Reinforcer on Human Cocaine Self-Administration," *Life Sciences* 55, no. 3 (1994): 179–87.

CHAPTER 6: DRUGS AND GUNS

1. National Household Survey on Drug Use and Health, 2010, http://www.samhsa.gov/data /NSDUH/2k10Results/Web/HTML/2k10Results.htm#7.1.5.

2. Christopher J. Mumola and Jennifer C. Karberg, U.S. Department of Justice, Office of Justice Programs, Bureau of Justice Statistics Special Report, Drug Use and Dependence, State and Federal Prisoners, 2004.

3. Ibid.

4. P. J. Goldstein, H. H. Brownstein, P. J. Ryan, and P. A. Bellucci, "Crack and Homicide in New York City: A Case Study in the Epidemiology of Violence," in Craig Reinarman and Harry G. Levine, eds., *Crack in America: Demon Drugs and Social Justice* (Berkeley: University of California Press, 1997), pp. 113–30.

5. S. R. Dube et al., "Childhood Abuse, Neglect, and Household Dysfunction and the Risk of Illicit Drug Use: The Adverse Childhood Experiences Study," *Pediatrics* 111, no. 3 (March 2003): 564–72, http://pediatrics.aappublications.org/content/111/3/564.long.

CHAPTER 7: CHOICES AND CHANCES

1. Anna Aizer and Joseph J. Doyle Jr., "Juvenile Incarceration and Adult Outcomes: Evidence from Randomly-Assigned Judges," National Bureau of Economic Research, February 2011.

2. U. Gatti, R. E. Tremblay, and F. Vitaro, "Iatrogenic Effect of Juvenile Justice," *Journal of Child Psychology and Psychiatry* 50 (2009): 991–98.

3. T. J. Dishion, F. McCord, and J. Poulin, "When Interventions Harm: Peer Groups and Problem Behavior," *American Psychologist* 54 (1999): 755–61.

4. Campaign for Youth Justice, "Critical Condition: African American Youth in the Criminal Justice System," September 25, 2008, p. 1, http://www.campaignforyouthjustice.org.

5. Ibid., pp. 16, 27.

CHAPTER 8: BASIC TRAINING

1. Jeffrey Haas, *The Assassination of Fred Hampton: How the FBI and the Chicago Police Murdered a Black Panther* (Chicago: Lawrence Hill Books, 2009).

2. R. Balko, "Overkill: The Rise of Paramilitary Police Raids in America," white paper, 2006.

3. Office of National Drug Control Policy, *National Drug Control Strategy: Data Supplement 2011* (2012), http://www.whitehouse.gov/sites/default/files/ondcp/policy-and-research/2011_data_supplement.pdf.

4. Craig Reinarman and Harry G. Levine, eds., *Crack in America: Demon Drugs and Social Justice* (Berkeley: University of California Press, 1997), p. 19.

5. Edith Fairman Cooper, *The Emergence of Crack Cocaine Abuse* (New York: Novinka Books, 2002), p. 49.

6. L. D. Johnston et al., *Monitoring the Future: National Survey Results on Drug Use, 1975–2011, vol. 1, Secondary School Students* (Ann Arbor: Institute for Social Research, University of Michigan, 2012).

CHAPTER 9: "HOME IS WHERE THE HATRED IS"

1. M. Daly and M. Wilson, "Competitiveness, Risk Taking, and Violence: The Young Male Syndrome," *Ethology and Sociobiology* 6 (1985): 59–73.

2. L. D. Johnston et al., *Monitoring the Future: National Survey Results on Drug Use, 1975–2011, vol. 1, Secondary School Students* (Ann Arbor: Institute for Social Research, University of Michigan, 2012).

3. Sudhir Venkatesh, *Gang Leader for a Day: A Rogue Sociologist Takes to the Streets* (New York: Penguin Press, 2008); Sudhir Venkatesh, *Off the Books: The Underground Economy of the Urban Poor* (Cambridge, MA: Harvard University Press, 2006).

4. Quoted in *Newsweek*, June 16, 1986.

5. Associated Press, "Browns Safety Dies of Cardiac Arrest," *New York Times*, June 28, 1986, http://www.nytimes.com/1986/06/28/sports/browns-safety-dies-of-cardiac-arrest.html.

6. Lynn Norment, "Charles Rangel: The Front-line General in the War on Drugs," *Ebony*, March 1989.

7. African American Members of the United States Congress: 1870–2008, *Congressional Record*, HR 5484, http://thomas.loc.gov/cgi-bin/bdquery/z?d099:H.R.5484.

8. U.S. Sentencing Commission, Report to the Congress: Cocaine and Federal Sentencing Policy, May 2007, p. 16.

CHAPTER 10: THE MAZE

1. Roberta Spalter-Roth, Olga V. Mayorova, and Jean H. Shin, "The Impact of Cross-Race Men-

toring for 'Ideal' and 'Alternative' PhD Careers in Sociology," American Sociological Association, Department of Research and Development, August 2011.

CHAPTER 12: STILL JUST A NIGGA

1. E. H. Williams, "Negro Cocaine Fiends Are a New Southern Menace," *New York Times,* February 8, 1914.

2. Ibid.

3. Ibid.

4. David Musto, *The American Disease: Origins of Narcotic Control,* expanded ed. (New York: Oxford University Press, 1987).

5. Ibid.

CHAPTER 13: THE BEHAVIOR OF HUMAN SUBJECTS

1. C. L. Hart et al., "Comparison of Intravenous Cocaethylene and Cocaine in Humans," *Psychopharmacology* 149 (2000): 153–62.

2. M. A. Nader and W. L. Woolverton, "Effects of Increasing the Magnitude of an Alternative Reinforcer on Drug Choice in a Discrete-Trials Choice Procedure," *Psychopharmacology* 105, no. 2 (1991): 169–74; M. A. Nader and W. L. Woolverton, "Effects of Increasing the Response Requirement on Choice Between Cocaine and Food in Rhesus Monkeys," *Psychopharmacology* 108 (1992): 295–300.

3. C. L. Hart and C. Ksir, *Drugs, Society, and Human Behavior,* 15th ed. (New York: McGraw-Hill, 2012).

4. L. R. Gerak, R. Galici, and C. P. France, "Self-Administration of Heroin and Cocaine in Morphine-Dependent and Morphine-Withdrawn Rhesus Monkeys," *Psychopharmacology* 204 (2009): 403–11.

5. D. K. Hatsukami and M. W. Fischman, "Crack Cocaine and Cocaine Hydrochloride: Are the Differences Myth or Reality?" *JAMA: The Journal of the American Medical Association* 276 (19) (1996): 1580–88.

6. C. L. Hart et al., "Alternative Reinforcers Differentially Modify Cocaine Self-Administration by Humans," *Behavioural Pharmacology* 11 (2000): 87–91.

7. Ibid.

8. S. T. Higgins et al., "Achieving Cocaine Abstinence with a Behavioral Approach," *American Journal of Psychiatry* 150, no. 5 (May 1993): 763–69.

9. M. Stitzer and N. Petry, "Contingency Management for Treatment of Substance Abuse," *Annual Review of Clinical Psychology* 2 (2006): 411–34.

10. K. Silverman et al., "A Reinforcement-Based Therapeutic Workplace for the Treatment of Drug Abuse: Six-Month Abstinence Outcomes," *Experimental and Clinical Psychopharmacology* 9, no. 1 (February 2001): 14–23.

CHAPTER 14: HITTING HOME

1. Bureau of Labor Statistics, U.S. Department of Labor.

CHAPTER 15: THE NEW CRACK

1. C. L. Hart and C. Ksir, *Drugs, Society, and Human Behavior,* 15th ed. (New York: McGraw-Hill, 2012).

2. J. A. Caldwell and J. L. Caldwell, "Fatigue in Military Aviation: An Overview of U.S. Military-Approved Pharmacological Countermeasures," *Aviation, Space and Environmental Medicine* 76, 7 (Suppl.) (2005): C39–51.

3. Remarks of Senator Barack Obama at Howard University Convocation, September 28, 2007.

4. Substance Abuse and Mental Health Services Administration, *Results from the 2011 National Survey on Drug Use and Health: Summary of National Findings,* NSDUH series H-44, HHS publication no. (SMA) 12-4713 (Rockville, MD: Substance Abuse and Mental Health Services Administration, 2012).

5. Fox Butterfield, "Home Drug-Making Laboratories Expose Children to Toxic Fallout," *New York Times,* February 23, 2004, http://www.nytimes.com/2004/02/23/us/home-drug-making-laboratories-expose-children-to-toxic-fallout.html?pagewanted=all&src=pm.

6.   Kate Zernike, "A Drug Scourge Creates Its Own Form of Orphan," *New York Times*, July 11,
      2005, https://www.nytimes.com/2005/07/11/national/11meth.html?pagewanted=2&sq=met
      hamphetamine%20scourge&st=cse&scp=1.

7.   S. L. Simon et al., "A Comparison of Patterns of Methamphetamine and Cocaine Use," *Journal
      of Addictive Diseases* 21 (2002): 35–44.

8.   C. L. Hart et al., "Is Cognitive Functioning Impaired in Methamphetamine Users? A Critical
      Review," *Neuropsychopharmacology* 37 (2012): 586–608.

9.   P. M. Thompson et al., "Structural Abnormalities in the Brains of Human Subjects Who Use
      Methamphetamine," *Journal of Neuroscience* 24 (2004): 6028–36.

10.   Hart et al., "Is Cognitive Functioning Impaired in Methamphetamine Users?"

11.   C. L. Hart et al., "Acute Physiological and Behavioral Effects of Intranasal Methamphetamine
      in Humans," *Neuropsychopharmacology* 33 (2008): 1847–55.

12.   A. Perez et al., "Residual Effects of Intranasal Methamphetamine on Sleep, Mood, and Perfor-
      mance," *Drug and Alcohol Dependence* 94 (2008): 258–62.

13.   B. A. Johnson et al., "Effects of Israddipine on Methamphetamine-Induced Changes in Atten-
      tional and Perceptual-Motor Skills of Cognition," *Psychopharmacology* 178 (2005): 296–302;
      B. A. Johnson et al., "Effects of Topiramate on Methamphetamine-Induced Changes in Atten-
      tional and Perceptual-Motor Skills of Cognition in Recently Abstinent Methamphetamine-
      Dependent Individuals," *Progress in Neuro-Psychopharmacology and Biological Psychiatry* 3
      (2007): 123–30; D. S. Harris et al., "The Bioavailability of Intranasal and Smoked Metham-
      phetamine," *Clinical Pharmacology and Therapeutics* 74 (2003): 475–86.

14.   M. G. Kirkpatrick et al., "Comparison of Intranasal Methamphetamine and *d*-amphetamine
      Self-Administration by Humans," *Addiction* 107 (2012): 783–91.

15.   C. L. Hart et al., "Alternative Reinforcers Differentially Modify Cocaine Self-Administration
      by Humans," *Behavioural Pharmacology* 11 (2000): 87–91.

16.   Hart et al., "Is Cognitive Functioning Impaired in Methamphetamine Users?"

17.   M. S. O'Brien and J. C. Anthony, "Extra-Medical Stimulant Dependence Among Recent Initi-
      ates," *Drug and Alcohol Dependence* 104 (2009): 147–55.

18.   Upton Sinclair, *I, Candidate for Governor: And How I Got Licked* (Berkeley: University of
      California Press, 1934), p. 109.

19.   Hart et al., "Is Cognitive Functioning Impaired in Methamphetamine Users?"

20.   G. Stix, "Meth Hype Could Undermine Good Medicine," *Scientific American*, December 27,
      2011.

CHAPTER 16: IN SEARCH OF SALVATION

1.   James Baldwin, *The Fire Next Time* (New York: Dial Press, 1963).

CHAPTER 17: DRUG POLICY BASED ON FACT, NOT FICTION

1.   http://www.fbi.gov/about-us/cjis/ucr/crime-in-the-u.s/2010/crime-in-the-u.s.–2010
      /persons-arrested.

2.   C. E. Hughes and A. Stevens, "A Resounding Success or a Disastrous Failure: Re-Examining
      the Interpretation of Evidence on the Portuguese Decriminalisation of Illicit Drugs," *Drug
      and Alcohol Review* 31 (2012): 101–113.

3.   C. Hart, "Remove the Knife and Heal the Wound: No More Crack/Powder Disparities,"
      Huffington Post, July 26, 2012, http://www.huffingtonpost.com/carl-l-hart/crack-cocaine
      -sentencing_b_1707105.html.